TRANSACTIONS

OF THE

AMERICAN PHILOSOPHICAL SOCIETY

HELD AT PHILADELPHIA
FOR PROMOTING USEFUL KNOWLEDGE

NEW SERIES—VOLUME 65, PART 5
1975

A GUIDE TO FRANCIS GALTON'S
ENGLISH MEN OF SCIENCE

VICTOR L. HILTS
Associate Professor of the History of Science, University of Wisconsin

THE AMERICAN PHILOSOPHICAL SOCIETY
INDEPENDENCE SQUARE
PHILADELPHIA

August, 1975

A GUIDE TO FRANCIS GALTON'S *ENGLISH MEN OF SCIENCE*

Victor L. Hilts

CONTENTS

INTRODUCTION

It is now almost exactly one hundred years since Francis Galton donned his best suit, strolled from his home near South Kensington to the meeting of the Fellows of the Royal Society of London and asked a number of those present to do something unprecedented.[1] Would they please, he asked, answer a simple questionnaire about themselves. What is the origin of your taste for science? How were you educated, and what were the merits and demerits of your education? Did religious beliefs hinder or help your scientific career? Perhaps because such questions had never been asked before of scientific men in such a scientific form, or perhaps because Francis Galton himself was an eminently respectable man of science, the return was not disappointing. There were about 190 men in all of whom Galton would have liked information; he was able to draw upon the answers of slightly over 100 scientists in writing the book which resulted from his questionnaire, *English Men of Science: Their Nature and Nurture.*

Although neglected until recently by psychologists, sociologists, and historians alike, Galton's *English Men of Science* deserves our attention for two reasons.[2] The first is the book's intrinsic importance in the history of the social and behavioral sciences: the theme of "nature and nurture" which afforded the subtitle for the book has been one of the most important and controversial distinctions in the social sciences ever since Galton himself saw its crucial significance. By way of definition Galton wrote in *English Men of Science* that "Nature is all that a man brings with himself into the world; nurture is every influence from without that affects him after birth."[3] The difference between nature and nurture is thus the difference between heredity and environment, between the congenital (or "innate") and the acquired, and ultimately between biology and culture. For at least ten years before the publication of *English Men of Science*, Galton himself had been convinced of the transcendent importance of nature as the chief cause of individual eminence in all fields of endeavor—science, literature, music, politics, athletics, and the military. *English Men of Science* was a case study which Galton hoped would substantiate his hereditarian thesis.

Overshadowing its role as a classic in the behavioral and social sciences, however, is the second reason Galton's book is now of importance. Because Galton's book was a product of the very first sociological questionnaire to concern itself with the scientific community, it is of great interest as a source of primary material on the personality and milieu of the Victorian scientist.

English Men of Science offers a vivid picture of the English scientist of one hundred years ago. Here we have published—though originally without identification—replies which a large number of English scientists gave to the very interesting questions which Galton asked of them. All of these scientists were eminent, most were found worthy of an obituary notice by the Royal Society, mention in the *Dictionary of National Biography (DNB)*, or reference in the *Encyclopaedia Britannica (EB)*; a few—like Darwin, Max-

[1] Francis Galton, *Memories of My Life* (London, 1908), p. 292.

[2] Two recent works that have spoken appreciatively of Galton's *English Men of Science* are Derek J. de Solla Price, *Little Science, Big Science* (New York, 1963), and Lewis S. Feuer, *The Scientific Intellectual* (New York, 1963). A reprint of *English Men of Science* was published in 1969 by Cass of London, with a brief introduction by Ruth Schwarz Cowan. For an earlier discussion, see Karl Pearson, *The Life, Letters, and Labours of Francis Galton* (3 v., Cambridge, 1914–1930) 2: pp. 134–156.

[3] Francis Galton, *English Men of Science: Their Nature and Nurture* (London, 1874), p. 9.

well, Cayley, Jevons—remain among the real giants of their respective sciences and are mentioned in every general history of science.

Although Galton did not identify his quotations in *English Men of Science*, fortunately, he did keep notes to those replies which enables one now to restore in many cases the personal element which was originally lost through the need to preserve statistical anonymity. Being familiar with Galton's writings and ways of doing things through having written a dissertation some years ago in which his work occupied a major role, it occurred to me that it might be possible to find somewhere in the Galton archives at University College, London, material pertaining to the actual questionnaires themselves. A grant from the Research Committee of the University of Wisconsin enabled me to visit London in hopes of finding such information. At first nothing materialized; there was no recollection on the part of any of the librarians of having ever seen what I was looking for. But very luckily that summer the Galton material was being transferred from its previous home with the Statistics Department to the University College archives themselves. One week before I was to leave London and return to Wisconsin, I was informed that some hitherto unknown material had been located when opening "Galton's desk." Among that material was a series of incomplete "Keys" (labeled by Galton as such) to *English Men of Science*. These "Keys" were sheets of paper on which Galton had given very short quotations from many scientists relating to particular questions of the questionnaire, with each "Key" relating to some section or chapter of the book. Two subsequent visits turned up other materials.[4] These materials have been made use of in preparing this "Guide to *English Men of Science*," the primary purpose of which is to identify as many as possible of the actual replies, thus affording historians an even better and more intimate look at the Victorian scientist.

One word should be said about the credibility of the material contained in *English Men of Science*. It might have been thought that Galton's work would be completely vitiated as an historical source by his great de-

sire to show the importance of hereditary ability among his scientists, and thus to revolve his analysis around the "nature versus nurture" dichotomy. Fortunately, however, Galton's *English Men of Science* suffers much less than might have been expected from Galton's own particular aim. In fact, Galton's special focus had one very important by-product for the historian and sociologist: it meant that his scientists included a considerable amount of information about their very early years. Galton's book also suffers less than might have been expected as a source of historical information for another reason: because the controversy with Alphonse de Candolle, which was the immediate cause of his study, made it necessary for Galton to deal with many factors other than heredity. In order to include all factors which had an influence on his scientists and determined their careers, Galton introduced the word "pre-efficients." Among the "pre-efficients" which he recognized—besides heredity—were home influences, professional influences, education, travel, and "accidents."

While Galton's *English Men of Science* has its undoubted deficiencies as a source of historical information, it has the one great advantage of being based upon inquiries actually directed to the persons being studied. As always one may doubt the veracity of statements made by individuals about themselves, especially when these statements concern events which happened many years ago in early life. Nevertheless, a direct answer to a direct question is certainly better as historical evidence than mere conjecture. The shrewd historian, recognizing that Galton's questionnaire was not exactly what he himself would have drawn up, and that no answer to the questionnaire can be accepted without a certain amount of caution, will make use of what is available and be thankful for that.

Chapter I of this "Guide" is a discussion of why Galton came to circulate his questionnaire. Chapter II examines the representativeness of the response which Galton received to the questionnaire by noting who did and who did not reply. Chapter III is a commentary upon Galton's discussion in *English Men of Science*. Chapter IV contains source material: an index table; identifications of replies which were quoted in *English Men of Science*; quotations from both *English Men of Science* and Galton's manuscripts, arranged by scientist.

One apology may perhaps be in order. It would have been useful to have included a reprint of *English Men of Science* with marginal identifications, and indeed this was at one time contemplated. However, the existence of a recent facsimile reprint of Galton's book and the current economics of the publishing profession have excluded this procedure.

---

[4] Whether or not there may still exist other material pertaining to *English Men of Science* remains an open question. Much of the Galton material passed through the hands of Karl Pearson when he was writing his biography of Galton, but there is no indication that Pearson had access to any material on *English Men of Science* not now available. It would seem, however, that Galton must have made extensive notes on two topics treated in *English Men of Science* for which no manuscript material has been located: the anthropometric measurements of mother and father, and the answers to the question about the possible deterrent effect of religion on science. It seems hard to believe that more of the questionnaires themselves have not survived in the families of the scientists who replied; they were rather impressive documents full of genealogical interest. Should any such questionnaire be discovered by scholars or others, the author of this "Guide" would be very happy to hear about them.

I. ORIGIN OF THE QUESTIONNAIRE

Since the primary use of *English Men of Science* is as a source of information on the lives of English scientists, it is appropriate that one describe the genesis of the book by first saying something about the Victorian scientist who was its author. Not only did Francis Galton qualify as a recipient of his own questionnaire, but he was also a member of a distinguished scientific genealogy. Like Charles Darwin, Francis Galton was a grandson of the eccentric physician and naturalist Erasmus Darwin.[1]

Galton's own career in fact bore some remarkable similarities to that of his more famous cousin. Like Charles Darwin, Francis Galton was religiously a freethinker. Like Charles Darwin, Galton had attended Cambridge but had not done exceptionally well there and could legitimately claim that, at least in science, he was to a large extent self-educated. Like Darwin also, Galton had spent a period traveling before settling down to scientific work. And finally, like Darwin, Galton had caught hold of a controversial idea which he realized could only be adequately proved or disproved, extended or restricted, by careful scientific investigation. Because Francis Galton, like Charles Darwin, was essentially a man of independence who chose freely to be a scientist, and because he had a theory which he believed could only be adequately discussed by scientists, Galton placed an extremely high value upon science. As he admitted in *English Men of Science*, "My bias has always been in favour of men of science, believing them to be especially manly, honest, and truthful. . . ."[2]

Galton made his first mark in science as a geographer. As *English Men of Science* itself bears witness, the field of geography offered important opportunities to several scientifically inclined individuals in the early and middle nineteenth century. Through geography one could enter the field of physics by way of navigation or meteorology; or one could enter natural history, geology, botany, or ethnology. Galton himself recognized the importance of geography among the specific influences which had affected several of his scientists. "Few men of scientific training have had opportunities of distant travel, but on those few their action has been very strong, especially as regards biologists and physicists."[3] In his own case, Galton was led via geography to the sciences of meteorology, ethnology, biology, and statistics, and was also led to take a wider view of the potentialities inherent in human nature itself. In *English Men of Science* he expressed this broadening aspect of travel, which was undoubtedly a stimulus to his thoughts about a new society based upon the new principles of eugenics, with characteristic charm:

Men are too apt to accept as an axiomatic law, not capable of further explanation, whatever they see recurring day after day without fail. So the dog in the backyard looks on the daily arrival of the postman, butcher, and baker, as so many elementary phenomena, not to be barked at or wondered about. Travel in distant countries, by unsettling these quasi-axiomatic ideas, restores to the educated man the freshness of childhood in observing new things and in seeking reasons for all he sees.[4]

In his chapter on the "Origin of Taste for Science" in *English Men of Science*, Galton includes material which suggests that in many cases a scientific career was determined by humble boyhood hobbies. So too, his own interest in travel developed from humble beginnings. First we have a story about how as a child Galton was encouraged by his father to make a short trip by himself—although unbeknownst to the young traveler the family servant was all the time following a discreet distance behind. Then we have the story about how he was sent to study chemistry in Germany while an undergraduate at Cambridge—but how once he got to Leibig's laboratory at Giessen, he discovered that he did not know enough German to benefit there and so wrote home, assuming paternal indulgence, that he had decided to take his money and see Constantinople.[5] Then we have a trip to Khartuum. Finally we have the trip of exploration and game-hunting which took Galton to tropical Southwest Africa in the early 1850's, and which provided his first real scientific activity.[6] For the purpose of this expedition Galton learned navigation, and during the expedition he charted unexplored areas and observed unfamiliar native customs. Once back in England, Galton published a book about his travels in Southwest Africa, became a member of the Athenaeum as a result, and begun to busy himself with work in the Geographical Society. Shortly he was elected secretary of the British Association, which gave him an opportunity to correspond with British scientists in all fields—an opportunity capitalized upon when circulating the questionnaire for *English Men of Science*. After his marriage, not long after his return from Africa, Galton made no more expeditions. After the late 1850's his new horizons were primarily intellectual ones, although every summer he and his wife traveled over the continent visiting various spas.

Having arrived home after his expedition in Africa, Galton thus settled down to the comfortable routine of a man with sufficient wealth to enjoy working in those activities which he considered particularly worth while. While some would have turned under such circumstances to politics or literature, Galton turned increasingly to science. Furthermore, his leisure allowed him the opportunity to follow his inclinations from interest to interest, without regard for any of those disciplinary boundaries which have so often proved confining. At first his most original work concerned meteorology—to which field he contributed the concept of the "anticyclone," and which undoubtedly familiarized him with

[1] Francis Galton, *English Men of Science*, p. 47.
[2] *Ibid.*, p. 148.
[3] *Ibid.*, p. 218

[4] *Ibid.*, pp. 218–219.
[5] Francis Galton, *Memories of My Life*, pp. 48–57.
[6] *Ibid.*, pp. 121–151.

the need for accurate statistics upon a large scale. Not long after, Galton turned to ethnological topics and began to speculate about improving the human race through a system of controlled marriage and breeding. The publication in 1859 of Darwin's *Origin of Species*, he said, encouraged him "to pursue many inquiries which had long interested me, and which clustered round the central topics of Heredity, and the possible improvement of the race."[7] The result of these speculations was Galton's hereditarian thesis: the thesis that hereditary talent and character is a determining factor for the individual and a valuable resource for the race.

Although Galton felt that his discovery about the importance of heredity was new, it was in fact not quite so unprecedented as he evidently believed. In his *Memories,* Galton wrote that "it seems hardly credible now that even the word heredity was then considered fanciful and unusual."[8] He added, "I was chaffed by a cultured friend for adopting it from the French." In spite of this, however, there were three aspects to Galton's hereditarian theses, and to some degree all three had been previously anticipated. These three aspects were: (1) that individual human characteristics are "innate" or "inborn"; (2) that innate characteristics are usually hereditary and can be transmitted from parents to offspring; (3) that the human race can be improved through selective breeding from individuals having the most desirable innate characteristics. Prior to Galton the belief that individual characteristics are innate was one of the chief tenants of the phrenologists, who were very influential in the early part of the nineteenth century. Before Galton, many medical writers, as well as phrenologists, believed that innate abilities are inheritable.[9] And finally, the idea that man could be biologically improved (the word more often used in the eighteenth and early nineteenth centuries was "perfected") by selective breeding had been discussed by various physicians, phrenologists, anthropolitis, and sanitarians.[10]

It is very doubtful that Galton himself was aware of the rather substantial history of his hereditarian ideas, and it is probably a good thing for his scientific development that he was not. Galton came to the hereditarian idea only after brooding deeply about racial differences, national character, and the nature of religion and human destiny. Possibly because he was not familiar with the prehistory of his own ideas, he saw all the more clearly that the hereditarian thesis needed scientific proof. Whereas others had suggested the importance of heredity in human affairs, no one before Galton took it upon himself to prove this importance, through the use of all means available, statistical and biological. It was this desire for scientific proof of his hereditarian thesis that sets Galton so much apart from earlier and contemporary thinkers with related ideas; and it was this desire which resulted, among other things, in the publication of *English Men of Science.*

Galton's first paper concerning his hereditarian ideas was published in 1865 under the title "Hereditary Talent and Character."[11] The paper gives evidence of the kind of ethnological speculations with which Galton had been concerned before focusing his attention upon the role of heredity. It also offered Galton's first attempt at a proof of his thesis; by a cursory analysis of several biographical dictionaries Galton concluded that there were a statistically significant number of relatives of eminent individuals who were themselves eminent— a fact which could be explained by the inheritance of ability in certain families. This lead was followed up by the publication in 1869 of *Hereditary Genius: An Inquiry into its Laws and Consequences.*

Again in *Hereditary Genius,* Galton offered evidence for his hereditarian thesis by an analysis of biographical material, but this time his analysis was much more extensive. In separate chapters he examined the biographies and the genealogies of persons eminent in such diverse fields as literature, politics, the army, and science. As with his earlier paper, but now in more detail, Galton found that very many eminent persons had relatives who were either eminent or nearly so in their own right. The Bach family of musicians and the Bernoulli family of mathematicians are only the most prominent examples. Galton realized in making his analysis that he was basing his judgment upon achievement and reputation (by taking his selection from biographical dictionaries) rather than ability it-

---

[7] *Ibid.,* p. 288.

[8] *Ibid.*

[9] In his notebook dated July, 1864 (in collection of University College, London) and in the resulting paper, Galton refers to some writings of the English author George Henry Lewes, who had neatly summarized some of the medical writings concerning human inheritance and who had argued that even "genius" is an inheritable trait. See George Henry Lewes, *Physiology of Common Life* (London, 1859).

[10] The long list of writers who had eugenic ideas is not generally known. One of the earliest of the modern eugenicists was the Spanish physician, Juan Huarte, whose work was translated into several languages, including an early English edition—Juan Huarte, *The Examination of Men's Wits,* 1594, translated by R. Carew with introduction by Carmen Rogers (Gainesville, Florida, 1959). Statements expressing the eugenic point of view are also found in two eighteenth-century writers—John Gregory, *A Comparative View of the State and Faculties of Man with those of the Animal World* (London, 1756), and more especially Charles A. Vandermonde, *Essai sur la manière de perfectionner l'espèce humaine* (Paris, 1756). Besides the phrenological writings, the eugenic position was

stated from an anthropological point of view in the early nineteenth century by William Lawrence in *Lectures on Comparative Anatomy, Physiology, Zoology, and the Natural History of Man* (London, 1819). Also contributing to the eugenic idea were the emerging medical fields of hygiene and public health; for the latter see brief discussion in V. L. Hilts, "William Farr (1807–1883) and the 'Human Unit,'" *Victorian Studies* 14 (1970): pp. 143–150. There were undoubtedly many eugenicists before Galton, and many of them seemed, like Galton, to believe in the complete novelty of their ideas.

[11] Francis Galton, "Hereditary Talent and Character," *Macmillians Magazine* 12 (1865): pp. 157–166, 318–327.

self, but he argued that in the cases which he was considering—which was that of persons of the highest eminence—"I see no reason to be dissatisfied with the conditions of accepting high reputation as a very fair test of high ability."[12] It seemed obvious enough to Galton that no man could become an outstanding success in any field without possessing related abilities, no matter what other advantages he might have.

There is very little to interest the modern reader of *Hereditary Genius* in the many genealogical pages included by Galton. More interesting, and much more historically important, was Galton's extension of his statistical method in the book. In the first chapters of *Hereditary Genius*, Galton showed how the curve of error could be used as a scale for measuring quantitatively the amount of ability; whereas the error curve had previously been used almost thirty years before by the Belgian statistician Adolphe Quetelet for measurements in physical anthropology, it had never previously been used in the field of mental measurements.[13] The use of the error curve for measuring ability was introduced by Galton in hopes of making his hereditarian thesis yet more precise: with its aid Galton might be able to show not only that ability is inherited, but just *how much* is inherited in particular cases.

Because of his use of the error law to measure hereditary ability, the main thrust of Galton's discussion in *Hereditary Genius* emphasized quantitative differences of ability. However, by dealing successively in different chapters with individuals eminent in diverse fields, Galton also gave the impression of believing in qualitative differences as well. It was quite obvious, for example, that Galton was speaking about different kinds, not different amounts, of ability when he contrasted in *Hereditary Genius* the character of "divines" in one chapter with that of "men of science" in another. The opposition between the character of the two kinds of men was, as Galton saw it, the opposition between the skeptic and the man of piety. Since this same opposition pervades Galton's discussion of the "independence" of scientists in *English Men of Science*, it may be well to give a quotation here:

In order that a man may be a contented sceptic of the most extreme type, he must have confidence in himself, that he is qualified to stand absolutely alone in the presence of the severest trials of life, and of the terrors of impending death. His nature must have sufficient self-assertion and stoicism to make him believe that he can act the whole of his part on earth without assistance. This is an ideal form of the most extreme scepticism, to which some few may nearly approach, but it is questionable if any have ever reached. On the other hand, the support of a stronger arm, and of a consoling voice, are absolute necessities to a man who has a religious disposition. He is conscious of

an incongruity in his nature, and of an instability in his disposition, and he knows his insufficiency to help himself.[14]

From Galton's implied admission of the existence of different kinds of hereditary ability, one is led naturally to the particular controversy which resulted in his publishing *English Men of Science*. The controversy is made more intelligible by asking some of the questions that might occur to a reader of *Hereditary Genius*. If there are differences in kinds of ability distinguishing individuals, do these innate differences explain, for example, why one person is a scientist and one person a theologian? Do divines become divines because of a hereditary "instability of disposition," while persons possessing the necessary hereditary talents, including independence of character, always wind up as eminent scientists? Is there, in other words, a kind of hereditarian predestination? Is "nature" everything and "nurture" nothing? In spite of his attempts at a scientific proof of his hereditarian thesis, it turned out that Galton was not immune from such objections as might seem contained in the above questions.

At the time Galton was collecting the material for *Hereditary Genius*, a Swiss biologist, Alphonse De Candolle, had already begun a work of a sociological character which was to give as much weight to "nurture" as Galton was giving to "nature."[15] De Candolle, himself, was well placed to appreciate the importance of different historical and social factors on scientific productivity because of the interesting position of his home city, Geneva, in the history of science. In the eighteenth century, Geneva had a far greater role in the history of science that might have been anticipated on the basis of its population alone. Was the explanation hereditary scientific ability? On the contrary, De Candolle thought that the ascendency of science in Geneva at that time was due primarily to the favorable social factors which characterized Switzerland. Lack of religious persecution and the existence of a group of intellectually oriented refugees, the Huguenots, were among the chief causes which De Candolle believed responsible for the flourishing of Genevese science.

In 1873 De Candolle published the results of his sociological study of the factors favoring scientific productivity in a volume entitled *Histoire des sciences et des savants depuis deux siècles*. The task which he set himself, as indicated by the title, was to investigate the social conditions which had been most favorable to science during the preceding two centuries. His method of attack was to examine the foreign membership in the scientific academies of London, Berlin, and Paris in order to discover which countries were most productive of scientific talent during the four particular years 1750, 1789, 1829, and 1869. Armed with this knowledge he

[12] Francis Galton, *Hereditary Genius: An Inquiry into its Laws and Consequences* (New York, 1870), p. 49.

[13] See comparison between Galton and Quetelet in V. L. Hilts, "Statistics and Social Science," in Giere and Westfall, *Foundations of Scientific Method: The Nineteenth Century* (Bloomington, Indiana, 1973), pp. 206–233.

[14] Francis Galton, *Hereditary Genius*, pp. 279–280.

[15] Alphonse De Candolle, *Histoire des Sciences et des Savants depuis deux siècles suivre d'autre études sur des sujet scientifiques* (Geneva, 1873).

then attempted to find the factors most nearly correlated with the flourishing of science. As a result some eighteen factors were isolated:

(1) A significant number of persons belonging to the rich or independent classes relative to those who have to work for a living and especially to those who have to work by manual labor; (2) A large number of individuals among the rich and independent classes who are contented with their situation, who have a fortune to administer, and who as a result are disposed to concern themselves with intellectual things of little remunerative value; (3) A tradition of intellectual culture which has been directed for several generations towards real things and correct ideas; (4) Immigration of respectable families who have a taste for intellectual things which offer little remuneration; (5) The existence of several families which have traditions favorable to science and to intellectual occupations of all kinds; (6) Well-organized primary, and especially secondary and higher education, which is independent of political parties or religions and which tends to stimulate scientific research and to encourage youths and their professors who are devoted to science; (7) Well-organized facilities for the different kinds of scientific work (libraries, observatories, laboratories, museums); (8) A public interested in truth and reality rather than imaginary or fictitious things; (9) Freedom to state and publish all opinions, at least on scientific topics, without experiencing any serious harm; (10) Public opinion favorable to science and scientists; (11) Freedom to engage in all professions, or not to engage in any, to travel, and to avoid all personal service except that which is undertaken voluntarily; (12) Religion which has little reliance upon authority; (13) Clergy who are friends of education among their own members and among the public; (14) Clergy not tied down or celibate; (15) Common use of one of the three principal languages, English, German, or French. Knowledge of these languages sufficiently extensive among the educated classes; (16) A small independent country or a union of small independent countries; (17) A geographical location with either a northern or temperate climate; (18) Proximity to advanced countries.[16]

One of the things discussed at length by De Candolle was the effect of religion upon science. Unlike Galton, who in *Hereditary Genius* had contrasted the personality of scientists and "divines," De Candolle emphasized the deterrent effect of religious dogmatism upon scientific productivity—thus shifting the ground from biology to sociology and history. Francis Galton himself gave a favorable summary of De Candolle's conclusions concerning science and religion in a review of *Histoire des sciences et des savants*.[17]

Alphonse De Candolle had given in his *Histoire des sciences et des savants* very substantial evidence that scientific productivity was dependent upon social and historical factors. Did this negate the position which Francis Galton had taken concerning the hereditary nature of scientific talent in *Hereditary Genius?* De Candolle himself certainly thought that this was the case and used Galton's hereditarian ideas—as Francis Galton described it—as a "foil" to set off his own con-

clusions.[18] Perhaps not implausibly De Candolle interpeted Galton to mean that hereditary abilities would develop themselves irrespective of social factors. Thus simply by showing variations which had occurred in the scientific productivity of different countries at different times, and by suggesting the social causes responsible for these variations, De Candolle had no difficulty in disposing of Galton's argument as he understood it.

Almost immediately upon receipt of De Candolle's *Histoire des sciences*, Galton entered into a correspondence with De Candolle; and he began to think about his own study of scientists.[19] The correspondence was extremely cordial, and in spite of the conflict in published views, a recognition seemed to develop on the part of both men that their two approaches were complementary rather than antagonistic. Galton assured De Candolle in the correspondence that he had *not* meant to imply that the inheritance of scientific ability led inevitably to scientific achievement. "I never said nor thought, that special aptitudes were inherited so strongly as to be irresistible, which seems to be the dogma you are pleased to ascribe to me and to repudiate," Galton wrote to De Candolle two weeks after having first received De Candolle's book.[20] Five months later Galton was able to report to De Candolle that the comments about the effect of religion upon science in *Histoire des sciences et des savants* had been quoted by Lyon Playfair in the debate about the future of science in Ireland and its relationship to Catholicism. Playfair, wrote Galton, "spoke with excellent effect and success, and I know that he derived at least that part of his argument from you, because I had myself directed his attention to your work previously as having a direct bearing on his then proposed speech."[21]

It was because of the challenge presented by De Candolle that Galton decided to study the role of hereditary ability among scientists. But whereas De Candolle analyzed the records of previous scientific generations, Galton accumulated fresh information about living scientists. There were several reasons for Galton's choice of material. First of all, Galton recognized the inherent difficulty of gleaning much specific information for his purpose from available records. It would be very difficult, for instance, to find published biographical material respecting similar talents possessed by deceased

16 *Ibid.*, pp. 196–197.
17 Francis Galton, "On the Causes Which Operate to Create Scientific Men," *Fortnightly Review* 13, N.S. (1873): pp. 348–349.

18 *Ibid.*, p. 346.
19 See letters in Karl Pearson, *The Life, Letters, and Labours of Francis Galton* 2: pp. 134–156. Also see G. de Morsier, "Correspondence inedité entre Alphonse de Candolle (1806–1893) et Francis Galton (1822–1911)," *Gesenerus* 29 (1972): pp. 129–160.
20 Francis Galton to A. De Candolle, December 27, 1872, quoted in Karl Pearson, *Life, Letters, and Labours of Francis Galton* 2: p. 139.
21 Francis Galton to A. De Condolle, May 7, 1873, quoted in Karl Pearson, *Life, Letters and Labours of Francis Galton* 2: p. 139.

scientists and their parents—even when such similarities may have existed and been the result of hereditary transmission. On the other hand, it was easy to ask living scientists about their parents. Secondly, Galton wanted his group of scientists to be as homogeneous among themselves as possible, and this objective was defeated when comparing scientists living at different times and in different countries. Lastly, as a good ethnologist and a sophisticated traveler, Galton was aware of the general difficulty of really understanding individuals from an alien setting.

Galton nowhere explains where he received the idea of using a questionnaire as a basis for his study, and the idea was by no means such an obvious one in 1873. However, as a member of the Statistical Society of London, Galton was undoubtedly familiar with statistical questionnaires, and he indicated in his *Memories* that he may have received help in his own questionnaire from a leading member of the Statistical Society, the eminent vital statistician, William Farr.[22] As an ethnologist, Galton was familiar with ethnological questionnaires. Furthermore, at almost the same time as he was preparing the questionnaire for *English Men of Science*, Galton was busily preparing another questionnaire, on the similarities and differences of identical and fraternal twins—a questionnaire which he also hoped would result in evidence of the importance of nature as opposed to nurture. Thus, while the idea of circulating a questionnaire was not such an obvious one in 1873 as it would be later, the method was not by itself precedent-setting. It is interesting to note that at almost the same time Charles Darwin was also circulating a questionnaire.[23]

At the end of *English Men of Science*, Galton gives a summary of the questionnaire which he used, and this obviates the necessity of describing it here. Once finished with the completed questionnaires, Galton evidently sent them back to their authors, and most have now vanished completely. One completed questionnaire does now exist in the University College archives: that filled out by the British Museum biologist Albert Günther. In addition, long excerpts from Charles Darwin's replies to Galton's questionnaire have long been publicly known, since they were published in Francis Darwin's *Life and Letters of Charles Darwin*. Because of the inherent interest of Darwin's replies, these excerpts are reproduced here (figs. 1–3).[24]

There is one part of Darwin's response to the questionnaire which may seem puzzling, and is deserving of explanation. This has to do with Darwin's remark

Education ?		
How taught?	I consider that all I have learnt of any value has been self-taught.
Conducive to or restrictive of habits of observation?	Restrictive of observation, being almost entirely classical.
Conducive to health or otherwise?	Yes.
Peculiar merits?	None whatever.
Chief omissions?	No mathematics or modern languages, nor any habits of observation or reasoning.
Has the religious creed taught in your youth had any deterrent effect on the freedom of your researches?	No.
Do your scientific tastes appear to have been innate?	Certainly innate.
Were they determined by any and what events?	My innate taste for natural history strongly confirmed and directed by the voyage in the *Beagle*.

FIG. 1. Excerpts from Darwin's Questionnaire.

that "I find it quite impossible to estimate my character by your degrees." This may seem puzzling because Galton himself does not discuss "degrees" of ability in *English Men of Science*. However, Galton did originally hope to rank every scientist with respect to several different qualities, and from this ranking to get an idea concerning the distribution of talent in the population. Thus the first part of his questionnaire (which is the only part not summarized in the appendix of *English Men of Science*) gave the following directions:

Qualifications to which no strictly defined meaning is attached, such as "large," "considerable," necessarily convey different ideas to different persons, and are unfit to be used statistically. I therefore beg to recommend in their place a notation of Classes which I proposed; justified, and employed in my *Hereditary Genius*, and which is explained in the following Table.
The successive graduations divide the Classes and start from Mediocrity as a zero point, and they proceed both upwards and downwards. They are calculated to show equal increments or decrements of the quality in question. The classes fill the intervals of, and are bounded by, the successive graduations.
The amount of the quality in question, indicated by each graduation, is that which exists at the limit which divides a select body of individuals from the residue of the population out of whom they have been selected, on the ground of their possessing that quality in an exceptional degree, either of excess or deficiency. The population is understood to be limited to individuals of the same age, and circumstances, and the severity of selection to be as is stated in the first column of the Table.

The table which Galton included, and to which he referred, was essentially the same table which he had included in *Hereditary Genius*.[25] In this table six classes above mediocrity (labeled in increasing order of distinctiveness: *A, B, C, D, E, F, G,* and *X*) and six classes below mediocrity (labeled in increasing order of distinctiveness: *a, b, c, d, e, f, g,* and *x*) were de-

[22] Francis Galton, *Memories of My Life*, p. 292.

[23] This questionnaire had to do with Darwin's study of the expression of emotions in man and animals. See R. B. Freeman and P. M. Gautry, "Charles Darwin's 'Queries about expression,'" *Bull. Brit. Museum (Nat. Hist.)* Hist. Ser. 4 (1972): pp. 207–219.

[24] Francis Darwin, *Life and Letters of Charles Darwin*, edited by his son (London, 1888), pp. 177–179.

[25] Francis Galton, *Hereditary Genius*, p. 34.

Question.	Yourself.			Your Father.	
Specify any interests that have been very actively pursued.	Science, and field sports to a passionate degree during youth.				
Religion?	Nominally to Church of England.			Nominally to Church of England.	
Politics?	Liberal or Radical.			Liberal.	
Health?	Good when young—bad for last 33 years.			Good throughout life, except from gout.	
Height, &c.?	Height?	Figure, &c.?	Measurement round inside of hat.	Height?	Figure, &c.?
	6 ft.	Spare, whilst young rather stout.	22¼ in.	6 ft. 2 in.	Very broad and corpulent.
	Colour of Hair?	Complexion?		Colour of Hair?	Complexion?
	Brown.	Rather sallow.		Brown.	Ruddy.
Temperament?	Somewhat nervous.			Sanguine.	
Energy of body, &c.?	Energy shown by much activity, and whilst I had health, power of resisting fatigue. I and one other man were alone able to fetch water for a large party of officers and sailors utterly prostrated. Some of my expeditions in S. America were adventurous. An early riser in the morning.			Great power of endurance although feeling much fatigue, as after consultations after long journeys; very active—not restless—very early riser, no travels. My father said his father suffered much from sense of fatigue, that he worked very hard.	

Fig. 2. Excerpts from Darwin's Questionnaire.

fined through the use of the curve of error. In order to do this, Galton took the probable error as a unit and defined the various classes by calculating the number of persons per million whose distinctiveness would place them, according to the law of error, less than one probable error from mediocrity (classes A and a), between one and two times the probable error from mediocrity (classes B and b) betweeen two and three times the probable error from mediocrity (classes C and c) and so on. According to this scheme persons in categories A and a were as abundant as 256,741 in one million, whereas persons in the other categories were increasingly rare. The number per million in each class was calculated by Galton as follows: 256,741 (A and a); 161,279 (B and b); 63,563 (C and c); 15,696 (D and d); 2,423 (E and e); 233 (F and f); 14 (G and g); and, finally, 1 (X and x). As a final note, Galton gave directions which would eliminate any references on his questionnaire to abilities possessed in only moderate degrees: "Wherever you consider the grade of the quality, about which a question is asked, to fall so near Mediocrity as to lie between C and c, do not make any entry at all." This meant that he was only concerned with individuals whose distinctiveness either above or below the average entitled them, at least, to be placed in categories D or d, respectively.

One is not surprised that Galton should have attempted to get a quantitative statement concerning de-grees of ability. Since the writing of *Hereditary Genius*, he had become increasingly entranced with the idea that, through the use of the law of error, quantitative measures of any ability could be constructed by considering the probable error to be the "statistical unit." In fact, much of a lecture which Galton delivered at the Royal Institution in February, 1874, which had his questionnaires as its subject, dealt with the possibility of dealing with his statistics in this sophisticated way.[26] However, it is not surprising either that few of his readers were willing to follow him on such advanced ground. Had Galton introduced a graph, his remarks might have been more intelligible, but he did not do so. By the time *English Men of Science* was ready to be published, Galton had to admit the frustration of his hopes for a truly quantitative science. The only vestige of his system of degrees occurs in a footnote to the questionnaire as reproduced in the appendix. There Galton wrote "I also omit the description of a notation I proposed to replace indefinite words such as 'large,' 'considerable,' because I have made no use of it in this volume." He added, "I have by no means abandoned

[26] Francis Galton, "Men of Science, Their Nature and Their Nurture," *Nature* 9 (1874): pp. 344-345. In this lecture Galton said, "The habit should . . . be encouraged in biographies, of giving copious illustrations which tend to rank a man among his contemporaries, in respect to every quality that is discussed in order to give data for appraising those qualities in terms of the Statistical Scale."

Energy of mind, &c.?	Shown by rigorous and long-continued work on same subject, as 20 years on the 'Origin of Species,' and 9 years on 'Cirripedia.'	Habitually very active mind—shown in conversation with a succession of people during the whole day.
Memory?	Memory very bad for dates, and for learning by rote ; but good in retaining a general or vague recollection of many facts.	Wonderful memory for dates. In old age he told a person, reading aloud to him a book only read in youth, the passages which were coming—knew the birthdays and death, &c., of all friends and acquaintances.
Studiousness?	Very studious, but not large acquirements.	Not very studious or mentally receptive, except for facts in conversation—great collector of anecdotes.
Independence of Judgment?	I think fairly independent ; but I can give no instances. I gave up common religious belief almost independently from my own reflections.	Free thinker in religious matters. Liberal, with rather a tendency to Toryism.
Originality or Eccentricity?	—— thinks this applies to me ; I do not think so—*i. e.*, as far as eccentricity. I suppose that I have shown originality in science, as I have made discoveries with regard to common objects.	Original character, had great personal influence and power of producing fear of himself in others. He kept his accounts with great care in a peculiar way, in a number of separate little books, without any general ledger.
Special talents?	None, except for business as evinced by keeping accounts, replies to correspondence, and investing money very well. Very methodical in all my habits.	Practical business—made a large fortune and incurred no losses.
Strongly marked mental peculiarities, bearing on scientific success, and not specified above?	Steadiness — great curiosity about facts and their meaning. Some love of the new and marvellous. —— N. B.—I find it quite impossible to estimate my character by your degrees.	Strong social affection and great sympathy in the pleasures of others. Sceptical as to new things. Curious as to facts. Great foresight. Not much public spirit—great generosity in giving money and assistance.

Fig. 3. Excerpts from Darwin's Questionnaire.

its advocacy, but have learned the necessity of explaining and exemplifying it in considerable detail before its merits and convenience are likely to become as generally recognized as I believe they deserve to be." [27]

Failure to produce a truly quantitative study, which would be a worthy successor of *Hereditary Genius*, was undoubtedly Galton's greatest disappointment in *English Men of Science*. Otherwise, his results seemed to him to confirm his stress upon the importance of "nature," and Galton was able to work with extraordinary rapidity in preparing his analysis of the returns. Charles Darwin had sent back his questionnaire in May, 1873. On February 27, 1874, Galton delivered a Friday evening lecture at the Royal Institution summarizing his results. By that time some "115 answers had already been received, of which 80 or 90 were full and minute replies." [28] It seems quite likely that by this time Galton had already prepared the "Keys" which are the basis of this *Guide*. The preface to *English Men of Science* itself was written in November, 1874.

Although Galton seemed satisfied with his results, not all of the first readers were convinced of his argu-

ments. An interesting, though somewhat curious, tract in opposition to both the procedures and results of *English Men of Science* made its appearance very shortly. This tract, written by a Mr. Francis Lloyd, was entitled "A Scientific View of Mr. Galton's Theories of Heredity" and was evidently envisaged as the first of a series dedicated to exposing fallacious reasoning in modern "science." What is interesting about Mr. Lloyd's criticism is that, at least from the positivistic point of view, its criticisms were entirely valid:

What distinguishes science from other forms of enquiry is the presence of method in every stage. A fact—and on facts science is built up—may be ever so true, and yet, if not susceptible of verification, utterly valueless for scientific purposes. . . . Science . . . demands that every particle of knowledge it accepts, be of such a description, that it can be subjected to varied tests, the application of which must not in any way shake its validity, as else it becomes worthless. Hence the fact, that science has hitherto only been applicable to the least complex of phenomena.

When, therefore, Mr. Galton proffers the confessions of a few individuals as a basis for exact conclusions, he errs as wildly as those of the Greeks who accepted oral traditions and travellers' stories as objective truth. However sincere the answers of his correspondents may have been, verification is out of the question. The experience of courts of law alone would prove that no man's evidence concerning himself is reliable. It would, indeed, be contrary to the first principle of human nature, if such were

[27] Francis Galton, *English Men of Science*, appendix.
[28] Francis Galton, "Men of Science, Their Nature and Their Nurture," *Nature* **9** (1874) : p. 344.

the case. To see ourselves as others see us would be the greatest of misfortunes.[29]

Lloyd had very little use for Galton's assumption that scientists would give honest answers, simply *because* they were scientists: "Mr. Galton attempts to escape reflections of this kind, which cannot but force themselves upon him, by asserting that men of science are more honest, straightforward, and manly than any other class." [30] "That such an opinion can be advanced on purely subjective grounds by a Fellow of the Royal Society," [31] Lloyd continued, "is a striking proof of the slight degree to which scientific thought has leavened even the leading minds amongst us."

There is no evidence that Galton ever responded to Mr. Lloyd's criticisms, nor was such a response necessary. No one was more aware than Galton himself that proof based upon statistics was a highly elusive business. While the results of *English Men of Science* were consistent with the hereditarian thesis, they did not by themselves really prove it. It was for just this reason that, in the remaining three decades of his life, Galton devoted an extraordinary amount of energy to finding additional evidence for his theory, clarifying its implications, and improving his statistical methodology. One result of these efforts was his publication in 1888 of his important book, *Natural Inheritance;* another was his discovery of the "correlation coefficient" as a statistical technique by which the amount of variability inherited from one generation to another could be given precisely and quantitatively.[32]

In the end *English Men of Science* was simply a way-station in Galton's own search for a truly scientific statement of his hereditarian thesis. In terms of his own development the book itself was not so important as either his previously published *Hereditary Genius* or his future works on heredity and statistics. But as a record of the replies given to Galton's questionnaire— the first ever distributed to scientists—Galton's *English Men of Science* has permanent importance.

II. WHO REPLIED (AND WHO DID NOT)

The period in which Galton's scientists lived was a transitional one for English science. In the early part of the nineteenth century there had been cries, vocalized especially by the mathematician Charles Babbage, that English science was in a "decline." [1] Babbage was particularly concerned that English science did not measure up to French science, which was at the time able to count an elite group of mathematicians, chemists, and biologists. Although Babbage's criticism was not entirely justified, since England in the early part of the nineteenth century was favored with at least Thomas Young, Humphry Davy, and Michael Faraday, still there was a certain truth to the complaint. While continental mathematics and physics had been developing at the hands of Paris mathematicians like Laplace, Lagrange, and Fourier, English mathematics was in the comparative doldrums in the first third of the century. Nor did England have many naturalists who could be put alongside of Cuvier and Humboldt. In spite of Priestley, Cavendish, and John Dalton, the new chemistry was known as the "French chemistry."

The failure of English science in the early part of the century actually ran deeper than could be inferred simply by comparing names and nationalities of the most eminent contemporary scientists. Above all, there was very little institutional development of science in England before the time of Galton's own scientists. During the French Revolution, the scientific and educational institutions of France had been reformed and science teaching at a very high level indeed was introduced at the École Polytechnique. In the early part of the century the German universities followed suit, and soon English chemists would be trooping to the laboratories of Liebig and Bunsen. Meanwhile, however, Oxford and Cambridge made only imperceptible motions away from a reliance upon the classics as the basis for a curriculum designed more to turn out gentlemen than scholars. If it had not been for the various dissenting academies of the eighteenth century, the Scottish universities at Glasgow and Edinburgh, and finally the University of London, there would have been almost no opportunity for a serious student of science to receive an English education in his subject anywhere in the British Isles in the earliest decades of the century—or for the young teacher of science to find himself an academic appointment.

Fortunately for those Englishmen interested in the welfare of native science, the dismal situation of the early part of the century was beginning to brighten by the time Galton addressed his questionnaire. Whereas for various reasons Paris science suffered a decline in the middle of the century, English science underwent a resurgence. Charles Darwin almost singlehandedly gave a new eminence to English biology, and Darwin was surrounded with other biologists of importance, like Wallace, Huxley, and Richard Owen. Similarly James Clerk Maxwell took over from the French mathematicians and the English experimentalist, Fara-

[29] Francis Lloyd, "A Scientific View of Mr. Galton's Theories of Heredity," *Modern "Science," No. 1* (London, 1876), pp. 38–39.

[30] *Ibid.,* p. 39.

[31] *Ibid.*

[32] Francis Galton, "Co-relations and their Measurement, Chiefly from Anthropometric Data," *Proceedings Roy. Soc.* 45 (1888): pp. 133–145.

[1] See brief discussion of the development of the scientific profession in England in Everett Mendelsohn, "The Emergence of Science as a Profession in Nineteenth-Century Europe," in Karl Hill, ed., *The Management of Scientists* (Boston, 1964), pp. 22–39. Also see D. S. L. Cardwell, *The Organization of Science in England. A Retrospect* (London, 1957), and Gordon W. Roderick, *The Emergence of a Scientific Society in England, 1800–1965* (London, 1967).

day, to create one of the most important theoretical advances of nineteenth-century physics: "Maxwell's theory" of electricity and magnetism. Even in some of the lesser sciences England was not doing badly. Whereas statistics in the early part of the century had been most advanced in France and Belgium, England could claim the lead by the time of Galton's questionnaire—even discounting Galton's own work. Chemistry, it is true, had been captured primarily by the Germans, but there were still some important English contributors, like Frankland and Perkins.

Slowly even the educational institutions were beginning to respond to science, although only under pressure. A year before Galton's questionnaire, the Royal Commission of 1872, the Devonshire Commission, had recommended the improvement of university science.[2] Galton was therefore able to conclude *English Men of Science* with the simultaneous observation that things had improved and plea that the "advocates" of science not "relax their efforts."

A great and salutary change has undoubtedly come over the feeling of the nation since the time when the present leading men of science were boys, for education was at that time conducted in the interests of the clergy, and was strongly opposed to science. It crushed the inquiring spirit, the love of observation, the pursuit of inductive studies, the habit of independent thought, and it protected classics and mathematics by giving them the monopoly of all prizes for intellectual work, such as scholarships, fellowships, church livings, canonaries, bishoprics, and the rest. This gigantic monopoly is yielding, but obstinately and slowly, and it is unlikely that the friends of science will be able, for many years to come, to relax their efforts in educational reform.[3]

The improvement which Galton professed to witness was due at least in part to the men whose replies are recorded in *English Men of Science*. Several of Galton's scientists were pioneers in the teaching of university science. Many, especially the chemist Lyon Playfair and Thomas Henry Huxley, were likewise engaged in the larger debate concerning the reformation of English education in the interests of science. Indeed, on one level Galton's book itself was a contribution to that debate.

The fact that Galton's scientists occupied a transitional role in the development of English science adds greatly to the interest of the replies found in *English Men of Science*. Why should his scientists, coming from an era when science was both uncommon and comparatively unrewarded, have taken up the subject. Was it an "innate" instinct—as Galton himself suggested? Or were there other factors at work? Whatever the reason, it is of more than usual interest to read about the motivations, working habits, and educational experiences of this generation of English scientists.

Because science was not yet a normal professional activity, it was necessary for Galton to take pains in his selection of individuals to be included in the distribution of the questionnaire. As he explains in the book, he decided to use two criteria in order to insure that each individual to whom he sent his questionnaire really deserved the distinction. First of all such an individual had to be a member of the Royal Society, and second, he had to distinguish himself scientifically in some other formal way.

To have been elected a Fellow of the Royal Society since the reform in the mode of election, introduced by Mr. Justice Grove nearly thirty years ago, is a real assay of scientific merit. Owing to various reasons, many excellent men of science of mature ages may not be Fellows, but those who bear that title cannot but be considered in some degree as entitled to the epithet of "scientific." I therefore look upon this fellowship as a "pass examination," so to speak, and from among the Fellows of the Royal Society I select those who have yet further qualifications. One of these is the fact of having earned a medal for scientific work; another, of having presided over a learned Society, or a section of the British Association; another, of having been elected on the council of the Royal Society; another of being a professor at some important college or university. These and a few similar signs of being appreciated by contemporary men of science, are the qualifications for which I have looked in selecting my list of typical scientific men.[4]

In 1847 Justice Grove's reform had restricted membership in the Royal Society to productive scientists, but there were still many members of the Royal Society in 1873 who could not meet Galton's stringent criteria. The membership of the Royal Society at the time of Galton's questionnaire was approximately five hundred, of which about forty were foreign members. Galton found that only about 190 individuals met his own requirements, although it appears he relaxed his criteria at least three times to include individuals who were not members of the Royal Society. The most eminent non-member included was the philosopher Herbert Spencer; the other two non-members were the statistician Robert H. Greg and the geologist John S. Henslow (for whom a posthumous reply was evidently submitted).

One can only conjecture why Galton relaxed his criteria to bring in the three names just mentioned. Probably a questionnaire was given to Spencer because of the latter's reputation as an evolutionary philosopher, and because Spencer had given Galton advice in the preparation of the questionnaire. Probably the questionnaire was sent to Greg because Galton knew that Greg was one of three brothers with somewhat similar interests, and therefore he hoped that Greg's reply would be of special interest with respect to the "nature-nurture" problem. Similar reasoning may have led to the posthumous reply for John S. Henslow. It may well be that the questionnaire was sent by Galton to

[2] D. S. L. Cardwell, *The Organization of Science in England*, pp. 92–98.
[3] Francis Galton, *English Men of Science*, pp. 259–260.
[4] *Ibid.*, pp. 3–4.

TABLE 1

NUMBER OF SCIENTISTS QUALIFYING FOR RECEIPT OF
THE QUESTIONNAIRE

1) Anthropoligists and Archaeologists	4
2) Astronomers	19
3) Biologists	44
4) Chemists	21
5) Engineers	15
6) Geographers and Meteorologists	7
7) Geologists	17
8) Mathematicians	13
9) Physicians	29
10) Physicists	12
11) Statisticians	10
12) Field unidentified	1
TOTAL	192

Henslow's son, the botanist George Henslow (not a member of the Royal Society), and that the latter—rather than answering in his own behalf—drew up a reply for his father.

With respect to those scientists who were members of the Royal Society, a record of the individuals selected for receipt of the questionnaire exists in Galton's annotated copy of the printed list of Fellows of the Royal Society as of the date November 30, 1872. Beside the names of 189 of these Fellows, Galton has placed a check mark, and this is clearly his working list of both names and addresses of individuals to whom the questionnaire should be submitted—although it is, of course, possible that the questionnaire did not in fact reach them all. He was undoubtedly helped in screening the total 535 names on the printed list by the fact that it also indicated recipients of medals given by the Royal Society, and gave professional positions and membership in various other scientific organizations in some cases. That the list of 189 Fellows checked by Galton really is the list of persons to whom the questionnaire was actually sent is confirmed by the fact that it is nearly identical to the list of scientists in a notebook mainly devoted to calculations concerning the "Head Circumference of Scientific Man." This latter notebook was used by Galton in his investigation of the relaitonship between head size and energy which is discussed in *English Men of Science*.

TABLE 2

MEMBERSHIP IN SCIENTIFIC SOCIETIES IN 1869

1) Astronomical Society	528
2) Mathematical Society	111
3) Geographical Society	2,150
4) Chemical Society	518
5) Geological Society	1,204
6) Botanical Society	2,420
7) Zoological Society	2,920
8) Statistical Society	371
9) Inst. of Civil Engineers	1,700

The method of selection was not, as Galton himself was fully aware, quite perfect. For instance, since Alfred Russell Wallace was not among the list of members of the Royal Society as of November 30, 1872, he was evidently not sent a questionnaire, even though Wallace had already been a recipient of the distinguished Royal Medal awarded by the Royal Society.[5] There were at least some cases in which Galton's selection procedure did not capture scientists of real eminence, even though they were on the Royal Society membership list—for instance, those chemists who were primarily industry-oriented, like William H. Perkin (creator of the synthetic dye industry) and Hugo Müller (an applied chemist engaged by the De la Rue Company). Necessarily, there were some scientists who were on the ascendency, and who would have been on Galton's list a few years later—for instance, the mathematician William Kingdon Clifford, and the future Lord Rayleigh.

In spite of its imperfections, Galton's list of those to whom the questionnaire was given is interesting as an attempt to identify members of the established English scientific community. Knowing the names of the scientists to whom the questionnaire was given makes it possible to see the proportions coming from the different scientific fields, and thus to compare the relative interest in the various sciences which existed in England at the beginning of the last quarter of the nineteenth century. Table 1 gives the distribution in terms of scientific fields of the 192 scientists on Galton's selected list. In compiling this list it has not, of course, been possible to take into account the fact that several of Galton's scientists contributed to more than one field, and that several of his scientists can only be uncomfortably classified in any scientific field at all—for instance, Earl Stanhope, an historian classified with the statisticians.[6] It is especially difficult to draw distinctions between some of the biologists and some of the physicians.

As a measure of just how select Galton's scientists were, it is interesting to compare table 1 with a list

[5] One can be reasonably sure that a scientist whose name appears on the November 30, 1872, membership list of the Royal Society, but was not checked by Galton on his copy of that list, was not sent a questionnaire. With respect to others it is more difficult to be sure that a questionnaire was not sent; however, there is no evidence in any of the manuscripts that Wallace was sent a questionnaire.

[6] The scientist whose field is unidentified is Christopher Rice Mansel Talbot, who had taken an honors degree in mathematics at Oxford, but who seems not to have published any scientific works. Talbot was a distinguished Welsh politician who had joined the Royal Society before the reforms of Justice Grove, and he may have been chosen by Galton because Galton was interested in the genealogy of the Talbot family. It is also conceivable, however, that C. R. M. Talbot was mistaken by Galton for Talbot of Malahide, who had presided over the Anthropology Section of the British Association. For all the other scientists the "Index Table" indicates what has been taken as the primary field for the purpose of this classification.

of membership in the different scientific societies at the same period (table 2). For this purpose one may use a census of membership in scientific societies which was taken in 1869, four years before Galton circulated his questionnaire.[7] In 1869 there was no Society devoted exclusively to physics.

Since Galton did not answer his own questionnaire, replies were possible from 191 scientists. In actuality, however, replies were received from only 104, including the physician Henry Holland, of whom Galton wrote in the Preface that he "had published his autobiography, but he gave me much help colloquially." Who did and who did not reply? Galton was aware that the reader might wonder whether the most eminent scientists responded, but he denied that the book was weighted towards mediocrity. A general check on the representativeness of those replying in terms of overall quality can be obtained by comparing the proportion of Royal Medal winners among the 191 scientists to whom the questionnaire was sent and the 104 who replied. Altogether some 42 of the 191 scientists had won at least one Royal Medal as indicated by the Royal Society membership list of November 30, 1872. This compares well with the 21 winners of the Royal Medal among the 104 respondents to the questionnaire. Although the percentage of Royal Medal winners among the 191 scientists (22.0 per cent) is slightly higher than among the 104 scientists (20.2 per cent), the difference is not statistically significant.

As would be expected, a considerable number of those who did not reply to Galton's questionnaire can be excused in terms of advanced age; many who fell technically within Galton's selected list were at the time of his questionnaire in the seventies or eighties. As a result, a relatively better response was received from among those who were elected to the Royal Society after Justice Grove's reform than before. The number of scientists who replied from each field—using Galton's own classification as used in *English Men of Science*—is given in table 3.

In order to judge the representativeness of Galton's sample of 104 names, it is necessary to identify the individuals field-by-field. Although the group who answered the questionnaire does seem representative of the group to whom it was sent, judged by percentage of Royal Medal winners, there are nevertheless differences in the adequacy with which replies were received when fields are taken into account. For simplicity one may take up the fields in the order in which Galton himself presents them in Chapter II of *English Men of Science*. As Galton points out, this order is given in accordance with the sections of the British Association for the Advancement of Science.

TABLE 3

NUMBER OF SCIENTISTS ANSWERING GALTON'S QUESTIONNAIRE

1) Physicists	6
2) Astronomers	7
3) Mathematicians	4
4) Geographers and Meteorologists	6
5) Chemists	12
6) Geologists	9
7) Botanists	9
8) Zoologists	23
9) Physicians	13
10) Statisticians	7
11) Engineers	8
TOTAL	104

1. MATHEMATICIANS AND PHYSICISTS

From the institutional point of view, British physics hardly existed before the time of Galton's questionnaire, as there were very few university or other professional opportunities open to the British physicist *per se* in the early or middle nineteenth century. At the very time Galton was making his survey this situation was beginning to change: the Clarendon Laboratory at Oxford opened in 1872; the Cavendish Laboratory at Cambridge began under James Clerk Maxwell in 1873; and the first professional society of physicists, the London Physical Society, was established in 1874. But all of these changes were too late to affect very much the composition of Galton's group of "physicists and mathematicians," which instead reflects the early lack of professional identity by including a heterogeneous composite of physical and mathematical scientists, astronomers, geographers, meteorologists, physical geologists, mathematicians, and physicists.

Besides the necessarily composite nature of Galton's "physicists and mathematicians," the character of this group in terms of quality seems to have been diluted by the failure of several of the most eminent names to reply to the questionnaire. Probably the foremost experimental physicists whose names are missing from the 104 replies are James Prescott Joule, John Tyndall, and Charles Wheatstone. Tyndall's omission is more surprising than that of either Joule or Wheatstone because Tyndall was one of Galton's own close scientific friends, and at one time the two had gone mountain-climbing together. One suspects that it may have been Tyndall whom Galton had in mind when he mused in his *Memories* about the curious fact that some of those whom he knew well and who had nothing to hide refused to answer his questionnaire.[8]

Among the missing mathematical physicists are George Gabriel Stokes, the Lucasion Professor at Cambridge, and William Thomson, the future Lord Kelvin. Two eminent mathematical physicists to whom Galton did not send his questionnaire were Peter Guthrie Tait,

---

[7] Leone Levi, "A Scientific Census," *Nature* 1 (1869): pp. 99–100.

[8] Francis Galton, *Memories of My Life*, pp. 292–293.

who co-authored with Thomson the most important British physical text of the nineteenth century, and John William Strutt, Third Baron Rayleigh. In Rayleigh's case the omission was necessitated by the fact that he was not elected to the Royal Society until June 12, 1873, the very year of Galton's questionnaire.

Turning to the mathematicians one also finds some noteworthy omissions among the replies, even though the institutional situation in mathematics was quite different from that in physics. In terms of both educational opportunities and professional advancement, mathematics was a much more highly institutionalized discipline than physics. Together with classics, mathematics was established as a separate "tripos" at Cambridge and, as Galton noted, it was success in mathematics and classics—not physics—which led to "fellowships, church livings, canonries, bishoprics, and the rest." But Galton did not send his questionnaire to the famous Cambridge tutor of mathematicians, John Edward Routh, and he did not receive replies from either the writer of mathematics texts, Isaac Todhunter, or the distinguished student of mathematical invariants, James Joseph Sylvester. William Kingdon Clifford, who was both a famous mathematician and one of the best-known English scientists because of his popular writings, was not sent a questionnaire since he was not elected to the Royal Society until June 4, 1874.

Since there were really very few top-flight English mathematicians and physicists, the failure of some of these few to respond necessarily affected Galton's sample of physicists and mathematicians. All told this group comprised some twenty-three names, as follows:

A. *Physicists*

1) Thomas Andrews (1813–1885). Professor of chemistry, Queen's College, Belfast. Researches on ozone; discovered that gases cannot be liquified by pressure when under a critical temperature. Fifty-one papers in *Royal Society Catalogue.*

2) George C. Foster (1835–1919), Professor of natural philosophy at Andrew's University, Glasgow, 1862. Professor of physics at University College, London. Papers on organic chemistry, physical chemistry, and electricity.

3) William Grove (1811–1896). Judge. Vice-president of the Royal Institution. Professor of expermiental philosophy at the London Institution. Author of *The Correlation of Physical Forces,* 1846, and inventor of the "Grove" battery.

4) James C. Maxwell (1831–1879). Professor of experimental physics at Cambridge. *Treatise on Electricity and Magnetism,* 1873. Research in electricity and magnetism, the kinetic theory of gases, and mathematics.

5) Robert Salisbury (1830–1903). Statesman. Research on electricity, mostly not published. "On Spectral Lines at Low Temperature." President of the British Association, 1894.

6) Balfour Stewart (1828–1887). Director of Kew Observatory. Discoverer of law of radiant heat. Author of *Treatise on Heat,* 1866.

B. *Astronomers*

1) Richard Carrington (1826–1875). Brewer. Proprietor of observatory at Red Hill, Reigate, Surrey. Observational astronomer.

2) Warren De La Rue (1815–1889). Stationer. Astronomer and chemist. Inventor of the Photoheliograph for solar photography. Twice president of the Chemical Society and president of the Royal Astronomical Society.

3) William Lassell (1799–1880). Brewer. Proprietor of observatory at Starfield, near Liverpool. Research on the satellites of Uranus and improvement of the reflecting telescope.

4) Robert Main (1808–1878). Assistant to Astronomer Royal for twenty-five years. Radcliffe Astronomer at Oxford.

5) Laurence Parsons (1840–1908), Earl of Rosse. Proprietor of Irish estates and observatory at Birr Castle, Parsonstown, King's Co., Ireland. "An Account of Observations of the Great Nebula in Orion, made at Birr Castle, with three-feet and six-feet telescopes."

6) Edward Sabine (1788–1883). General. President of the Royal Society 1861–1871, and general secretary of British Association for Advancement of Science. Fifteen papers of "Contributions to Terrestrial Magnetism" in *Philosophical Transactions.*

7) Piazzi Smyth (1819–1900). Astronomer Royal for Scotland. One hundred papers in Royal Society's *Catalogue for Scientific Papers.*

C. *Mathematicians*

1) Arthur Cayley (1821–1895). Sadlerian Professor of pure mathematics at Cambridge. Research in theory of determinants and invariants. *Collected Mathematical Papers.*

2) Thomas Hirst (1830–1892). Director of naval studies at Royal Naval College, Greenwich. President of London Mathematical Society, 1872–1874. Research on correlation of planes and correlation of spaces of three dimensions.

3) Archibald Smith (1813–1872). Professor of mathematics at Cambridge. Founder of *Cambridge Mathematical Journal.* "New mode of correcting the quadrantal deviation of compass," 1861; with Evans, "The Manual for the Deviation of the Compass in Iron Ships."

4) William Spottiswoode (1825–1883). Printer. President of the Royal Society, 1878–1883. *Elementary Theorems Relating to Determinants,* 1851; *Polarisation of Light,* 1874. Ninety-nine papers in *Royal Society Catalogue of Scientific Papers.*

D. *Physical Geographers and Meteorologists*

1) George Back (1796–1878). Admiral. Author of *Narrative of the Arctic Land Expedition to the Mouth of the Great Fish River in 1833, 1834, and 1835.*

2) Frederick J. Evans (1815–1885). Captain. Superintendent of the compass department of the navy. Hydrographer to the admiralty. Research on terrestrial magnetism and magnetism in ships.

3) Sherard Osborn (1822–1875). Rear-admiral. *Last Voyage and Fate of Sir John Franklin,* 1865.

4) Samuel Haughton (1821–1897). Professor of geology at the University of Dublin. Editor of the *Dublin Quarterly Journal of Science.* Research on applied mathematics, physical geology, and meteorology.

5) Robert H. Scott (1833–1916). Civil engineer. Editor of H. V. Dove, *The Law of Storms considered in connection with the ordinary movements of the atmosphere,* 1862; *A Handbook of olumetric Analysis,* 1862. Treasurer of the Royal Society Club for seventeen years.

6) Richard Strachey (1817–1908). Lieutenant-general, Royal (Bengal) Engineers. Himalayan explorer.

2. CHEMISTS

Although to a lesser extent than with the physicists and mathematicians, Galton's sample of 104 names also omits some of the most important English chemists. It has already been noted that Galton's method of selection led to his overlooking two important industrial chemists, William H. Perkin and Hugo Müller. Among other missing names are Benjamin Brodie, 2nd, Edward Frankland, George Liveing, William Crookes, Johnstone Stoney, and, among the coming generation of chemists, the young James Dewar, who was not elected to the Royal Society until 1877.

Altogether it appears that some twelve chemists are known to have answered the questionnaire. Not surprisingly, considering the German lead in chemistry during the middle of the nineteenth century, two of these English chemists were in fact transplanted Germans, Henry Debus and Jochn C. Voelcker. In addition, Galton received a reply from Thomas Andrews, who was a professor of chemistry, but was classified by Galton—undoubtedly because of his work on the liquefaction of gases by pressure—among the physicists. Of of these twelve chemists about half were professors of chemistry, although only two were associated with either Oxford or Cambridge. Two of Galton's chemists were at one time associated with the University of London, one was currently at Royal Institution, one was at the University of Manchester, one was at the Royal Naval College. One of his chemists had been a professor at the Royal Agricultural College. The twelve chemists were as follows:

1) George B. Buckton (1818–1905). Research assistant of A. W. Hoffman at Royal College of Chemistry. Discoverer and isolator of mercuric methyl. Also an entomologist.

2) Henry Debus (1824–1916). Professor of chemistry at Royal Naval College.

3) John Hall Gladstone (1827–1902). Lecturer on chemistry at St. Thomas's Hospital, 1850–1852, and Fullerian Professor of chemistry at the Royal Institution, 1874–1877. Research on the refraction and dispersion of liquids. Pioneer in technical education.

4) Augustus Harcourt (1834–1919). Tutor of Christ's Church, Oxford. Research on the law of mass action.

5) John B. Lawes (1814–1900). Developer of phosphate fertilizer industry and originator of Rothampsted studies in experimental agriculture.

6) William H. Miller (1801–1880). Professor of mineralogy at Cambridge. Research on crystallography.

7) Lyon Playfair (1818–1898). Professor of chemistry at London 1858–1869, and secretary of the Department of Science and Art from 1853 to 1858. Member of parliament representing Edinburgh University, 1868–1885.

8) Henry E. Roscoe (1833–1915). Professor of chemistry at Owens College, Manchester. Research concerning the effect of light on chemical reactions.

9) John Stenhouse (1809–1880). Private commerical chemist in London. Research on dyes,

sugar manufacture, glue, and organic chemistry. Author of over one hundred papers listed in *Royal Society Catalogue.*

10) John C. Voelcker (1822–1884). Professor of chemistry at the Royal Agricultural College, Cirencester, 1849–1863. Consulting chemist in London from 1863.

11) Alexander W. Williamson (1824–1904). Professor of chemistry at University College, London. Research on ether and theory of ionic dissociation. Interest in chemical manufacturing processes.

12) Philip James Yorke (1799–1874). President of Chemical Society in 1853. Research on the solution of metallic lead by water.

3. GEOLOGISTS

By the middle of the nineteenth century the pioneer era of English geology was fast disappearing. Among the names of this era were William Buckland, Adam Sedgwick, William Conybeare, Roderick Murchison, Henry Thomas De la Beche, John S. Henslow, Charles Lyell, and John Phillips. By a quirk of irony four of these men—Sedgwick, Murchison, Lyell, and Phillips—died at almost the very time Galton was circulating his questionnaire, after having lived to ripe old ages that ranged between seventy-four and eighty-eight. In his preface Galton noted the death of Professor Phillips, but indicated that Phillips had submitted a nearly complete return. Charles Lyell, who was already blind, died four months after Galton wrote his preface, understandably without having completed the questionnaire. Sedgwick and Murchison both died just barely too soon to have found their place in Galton's stastistics. As has been noted, however, Galton seems to have found someone to complete a questionnaire posthumously for John S. Henslow, who had died over a dozen years earlier.

Although Galton missed some of the pioneer geologists, it seems that he did get a fairly representative return from among the next generation. Probably the most eminent younger geologist who was not included on Galton's list was Archibald Geike (1835–1924). The representativeness of Galton's sample is indicated by the fact that it included both university geologists and geologists associated with the Geological Survey. At the time of Galton's questionnaire the Geological Survey was housed in its research building on Jermyn Street, along with the Museum of Practical Geology, and the Royal School of Mines. Among Galton's geologists three (Warington Smyth, Ramsay, and Phillips) were connected at one time with the Geological Survey; Smyth also took up a lectureship at the School of Mines when it was founded in 1851.

Altogether there were nine geologists who replied to Galton's questionnaire. In addition, Galton received a reply from Samuel Haughton, who was a professor of geology at Dublin, but who had many interests and was included among the physicists. Because Henslow gave most of his attention to botany in the years after 1830, he has been included with the botanists. The nine geologists were as follows:

1) David T. Ansted (1814–1880). Professor of geology at King's College, London, and lecturer at Addiscombe and the Civil Engineering College at Putney. Editor of the *Quarterly Journal of the Geological Society. The Ancient World: or, Picturesque Sketches of Creation,* 1847; *Elementary Course of Geology, Mineralogy, and Physical Geography,* 1850; *The Great Stone Book of Nature,* 1863; *Applications of Geology to the Arts and Manufactures,* 1865. Interest in practical aspects of geology, including mining and water supply.

2) John Evans (1823–1908). Archaeologist and partner in a paper-manufacturing firm. *The Ancient Stone Implements, Weapons, and Ornaments, of Great Britain,* 1872; *The Ancient Bronze Implements, Weapons and Ornaments of Great Britain and Ireland,* 1881.

3) David Forbes (1828–1876). Mineralogist and metallurgist. Partner in a Birmingham nickel-smelting firm and foreign secretary of the Iron and Steel Institute. Research in microscopic petrology.

4) T. Rupert Jones (1819–1911). Geologist and paleontologist. Professor at the Royal Military College, Sandhurst. Editor of *Quarterly Journal of the Geological Society. A Monograph of the Entomostraca of the cretaceous formations of England,* 1849; *Manual of the Natural History, Geology and Physics of Greenland and the Neighboring Regions,* 1875; *Catalogue of the Fossil Foraminifera in the British Museum,* 1882.

5) N. Story-Maskelyne (1823–1911). Mineralogist. Professor of mineralogy at Oxford. Keeper of the mineral collection at the British Museum. Member of parliament. Research on meteorites and crystallography.

6) John Phillips (1800–1874). Reader of geology at Oxford from 1856 to 1874. *Illustrations of the Geology of Yorkshire: or, a description of the strata and organic remains of the Yorkshire Coast,* 1829–1836; *Life on Earth, its origin and succession,* 1860.

7) Joseph Prestwich (1812–1896). Stratigrapher, wine merchant and after Phillips's death the reader in geology at Oxford. *A Geological Inquiry respecting the Water-bearing Strata of the*

country around London, 1851; *Notes on Some further discoveries of Flint implements in beds of post-Pliocene gravel and clay*, 1861; *Geology, chemical, physical, and stratigraphical*, 1886–1888.

8) Andrew C. Ramsay (1814–1891). Director-general of the Geological Survey, lecturer in the Schools of Mines and professor of geology at London from 1847 to 1852. *A Descriptive Catalogue of the Rock Specimens in the Museum of Practical Geology*, 1858; *The Geology of North Wales*, 1866; *Passages in the History of Geology*, 1848.

9) Warington Smyth (1817–1890). Mineralogist and lecturer at the School of Mines. *A Catalogue of the Mineral Collections in the Museum of Practical Geology*, 1864. *A Treatise on Coal and Coal Mining*, 1866; *An Account of the Mining Academies of Saxony and Hungary*, 1846.

4. BIOLOGISTS

Galton received more replies from biologists than from any other group of scientists. Putting together the botanists, zoologists, and physicians nearly one-half of the total replies came from the biologists. Contrary to the case with the physicists and mathematicians, and to a lesser extent with the chemists, many of the foremost biological scientists of the time seem to have responded to the questionnaire. Those who replied included Darwin, Owen, and Huxley. Among the missing are the names of Alfred Russell Wallace and Henry W. Bates, to neither of whom the questionnaire had been sent, and William Jenner and Joseph Lister, both of whom did have an opportunity to reply.

The list of biologists who answered Galton's questionnaire accurately reflects the extraordinary importance of medicine as a stimulus to biology in the middle of the nineteenth century. Galton himself seems to have drawn the dividing line between the physicians and other biologists, not in terms of either education or research interests, but rather in terms of whether or not a particular scientist was engaged in actual practice. It may be added that consultation of the *Dictionary of National Biography* and other sources indicates that most of the physicians and surgeons seem not only to have engaged in practice, but to have done so very profitably.

Among Galton's biologists were representatives of several universities, although university positions would have been much less numerous if it had not been for the medical schools. Those who became professors at the Scottish universities were three: Allman (Edinburgh, natural history); J. H. Balfour (Glasgow, botany); and Allen Thomson (Glasgow, anatomy). There was only one biologist who was a professor at Ox-

ford (J. Burdon-Sanderson, medicine), and one each from Dublin (Stokes, medicine), and Owens College, Manchester (Williamson, natural history). Primarily because of their medical schools, University College, London, and King's College, London, offered professorships to several of Galton's biologists. There were three of Galton's biologists who eventually held positions at Cambridge: Newton (zoology and comparative anatomy); Babington (botany); and Humphry (medicine).

For those biologists who were not engaged in medicine, greater professional opportunities than were found in the universities were afforded by collections such as could be found at the British Museum, the Linnean Society, Kew Gardens, and the Hunterian Museum of the Royal College of Surgeons. Because the nineteenth century was a period of great collection building in English biology, some of it aided by the existence of numerous English voyages of scientific discovery and some of it simply representing the interests of private individuals, several of Galton's scientists were intimately involved in collecting and in making use of various collections. George Bentham, the nephew of the philosopher Jeremy Bentham, had a private collection which he donated to Kew Gardens with the proviso that he would be able to utilize its resources there. Richard Owen evidently owed some of his knowledge of comparative anatomy to his opportunities while preparing a catalog of the Hunterian Collection of the Royal College of Surgeons. Philip Lutley Sclater began collecting birds while still an Oxford undergraduate; eventually his collection of 8,824 specimens was acquired by the Natural History Museum. The conchologist Jeffreys brought up 71 new species in one haul while dredging at the depth of 994 fathoms off the coast of Ireland in H.M.S. *Porcupine*. Joseph Dalton Hooker, who worked in collaboration with Bentham at Kew Gardens on the great *Genera Plantarum* and the geographical distribution of species, drew up the botanical questions which guided the collecting of the *Challenger* expedition. In fact there might have been still other scientists connected with the *Challenger* expedition who might have replied to Galton's questionnaire except for the fact that the expedition left England in December, 1872, and did not return until two years after the publication of Galton's *English Men of Science*.

One of the centers of systematic collecting in the middle of the nineteenth century was the British Museum, and at least three of Galton's biologists were closely identified with work at the Museum. One of these was Richard Owen, who was superintendent of the natural history department of the British Museum before the establishment of the separate Natural History at South Kensington. Another was John Edward Gray, who was keeper of the zoological department of the British Museum; although Gray was seventy-four years' old and had already suffered a stroke, he answered Galton's questionnaire in some detail. A third representative of the British Museum staff was the German-born botanist Albert Charles L. G. Günther. Günther had received a Ph.D. from Tübingen in theology and had studied with the famous German biologist Johannes Müller. "Every collection of note made by explorers all over the world came to him at least for the fishes, frogs, and reptiles, and occasionally for the birds and mammals," it is written of Günther in his Royal Society obituary notice. What Günther wrote in answer to Galton's questionnaire seems to confirm this: "From 1858 to 1871 examined and named 40,000 examples, described 8000 species, wrote 6000 pages of print and corresponded also." Günther is of particular interest since his completed questionnaire is the only such questionnaire now extant in the University College archives.

Altogether there were forty-five biologists who responded:

A. *Botanists*

1) Charles Babington (1808–1895). Professor of botany at Cambridge. Research in botany and archaeology. *Manual of British Botany*, 1843.

2) John H. Balfour (1808–1884). Professor of botany at Glasgow. Editor of *Annals of Natural History. Outlines of Botany*, 1854.

3) John Ball (1818–1889). Member of parliament. First president of the Alpine Club. *Alpine Guide*, 1863–1868.

4) James Bateman (1811–1897). Horticulturist and expert on orchids. *A Second Century of Orchidaceous Plants*, 1867; *Monograph of Odontoglossum*, 1864–1874.

5) George Bentham (1800–1884). Researcher in Botanic Gardens at Kew. *Outline of a system of Logic;* 1827; *Genera Plantarum*, 1883.

6) Frederick J. Currey (1819–1881). Secretary of the Linnean Society. Research on fungi. Translator of Hofmeister's *On the Higher Cryptogamia.*

7) John S. Henslow (1796–1861). Professor of botany at Cambridge. *A Catalogue of British Plants*, 1829; *Principles of Descriptive and Physiological Botany*, 1835.

8) Joseph D. Hooker (1817–1911). President of the Royal Society (1873–1878). Director of Kew Gardens. *Outlines of the Distribution of Arctic Plants*, 1862; with G. Bentham, *Genera Plantarum*, 1862–1883.

9) John Miers (1789–1879). Developer of mineral resources in Chile and engineer in South America. *Travels in Chile and La Plata*, 1825; *Illustrations of South American Plants*, 1850–1857.

B. *Zoologists*

1) George J. Allman (1812–1898). Regius Professor of natural history at Edinburgh. Research on the classification and morphology of the coelenterata and polyzoa.

2) William B. Carpenter (1813–1885). Physiologlist and naturalist. Registrar of University of London. *The Microscope and its Revelations*, 1856; *Principles of General and Comparative Physiology*, 1852; *Principles of Mental Physiology*, 1874.

3) Henry J. Carter (1813–1895). Surgeon-Major in East India Company. Many publications on Indian geology and biology. Research on sponges.

4) Thomas Cobbold (1828–1886). Private practice in London. Swiney Lecturer on geology at the British Museum. Professor of botany (and subsequently helminthology) at the Royal Veterinary College. Research on human and animal parasitic worms.

5) Charles Darwin (1809–1882). No professional position. *Origin of Species*, 1859; *Descent of Man*, 1871.

6) Philip Grey-Egerton (1806–1881). Member of parliament. Publisher of several catalogues of fossil fishes.

7) William Flower (1831–1899). Director of Natural History Museum, London, and Hunterian Professor of comparative anatomy and physiology at the College of Surgeons. *Osteology of the Mammalia*, 1870.

8) John Edward Gray (1800–1875). Keeper of the zoological department of the British Museum. *Synopsis of British Molluscs*, 1852.

9) Albert Charles L. G. Günther (1830–1914). Staff of the British Museum and keeper of the zoological department 1875–1895. *Geographical Distribution of Reptiles*, 1858.

10) Thomas H. Huxley (1825–1895). Lecturer in the Royal College of Mines, Hunterian Professor at Royal College of Surgeons and Fullerian Professor at the Royal Institution. *Oceanic Hydrozoa*, 1859; *Theory of the Vertebrate Skull*, 1859; *Man's Place in Nature*, 1863.

11) John G. Jeffreys (1809–1885). Practicing lawyer. Researches in conchology. *British Conchology*, 1862–1869.

12) John Lubbock (1834–1913). Banker. *Prehistoric Times*, 1865; *The Origin and Metamorphoses of Insects*, 1873.

13) George Mivart (1827–1900). Lecturer on comparative anatomy, St. Mary's Hospital, London. Writings on comparative anatomy and in opposition to theory of natural selection. *Man and Apes*, 1873; *The Origin of Human Reason*, 1889.

14) Alfred Newton (1829–1907). Professor of zoology and comparative anatomy at Cambridge. *Dictionary of Birds*, 1893–1896.

15) Richard Owen (1804–1892). Hunterian Professor in the Royal College of Surgeons and superintendent of natural history in the British Museum. *Comparative Anatomy and Physiology of Vertebrates*, 1866–1868; *History of British Fossile Reptiles*, 1849–1884.

16) William K. Parker (1823–1890). Hunterian Professor of comparative anatomy at the Royal College of Surgeons. Research on the Foraminifera and on the anatomy of the vertebrate skull.

17) Edmund A. Parkes (1819–1876). Professor of hygiene at the Army Medical College. Research in military hygiene and on the physiology of exercise.

18) John Burdon-Sanderson (1828–1905). Jodrell Professor of physiology at University College, London, 1874–1882, professor of physiology at Oxford and Regius Professor of medicine at Oxford, 1895–1904. Research on the relationship between micro-organisms and disease and on electrical phenomena of plants.

19) Philip L. Sclater (1829–1913). Secretary to the Zoological Society. Editor of *The Ibis*, journal of the British Ornithologists Union. Over 1,200 papers on birds and mammals. Address on "The State of our Knowledge of Zoological Geography."

20) Herbert Spencer (1820–1903). Editor, writer, and philosopher. *Social Statics*, 1850; *Principles of Psychology*, 1855; *Synthetic Philosophy*, 1860–1896.

21) William Stokes (1804–1878). Regius Professor of medicine at Dublin. Early work on the stethoscope. *A Treatise on the Diagnosis and Treatment of Diseases of the Chest*, 1837.

22) Allen Thomson (1809–1884). Professor of anatomy at Glasgow. Research in embryology.

23) William C. Williamson (1816–1895). Professor of natural history, anatomy, and physiology at Owens College, Manchester. Research in paleobotany and on fossil fishes.

C. *Physicians and Surgeons*

1) James Alderson (1794–1882). President of the Royal College of Physicians. Consulting physician to St. Mary's Hospital, Paddington. *Diseases of the Stomach and Alimentary Canal*, 1847.

2) Thomas G. Balfour (1813–1891). Deputy inspector-general of the army medical department. Publications on army medical statistics, including *Statistics of the British Army*.

3) Henry Charlton Bastian (1837–1915). Professor of clinical medicine at University College, London. Research on nerves, brain, and spinal cord. *Brain as an Organ of Mind*, 1880.

4) William Bowman (1816–1892). Private practice and professor of physiology and general and morbid anatomy at King's College, London. Research in the structure of the eye and kidney and practice as an ophthalmic surgeon. *Lectures . . . on the Eye*, 1849.

5) Charles Brooke (1804–1879). Member of surgical staff of Metropolitan Free Hospital and Westminster Hospital. Inventor of the "bead suture" for deep wounds.

6) Arthur Farre (1811–1887). Private practice and professor of obstetric medicine at King's College, London. Obstetric surgeon and author of "The Uterus and its Appendages" in Todd's *Cyclopaedia of Anatomy and Physiology*, 1858.

7) William Fergusson (1808–1877). Professor of surgery at King's College Hospital and practicing surgeon. *System of Practical Surgery*, 1870.

8) Wilson Fox (1831–1887). Holme Professor of clinical medicine at University College, London. *Diseases of the Stomach*, 1872.

9) Henry Holland (1788–1873). Private practice. *Medical Notes and Reflections*, 1839; *Essays on Scientific and other subjects contributed to the "Edinburgh" and "Quarterly" Reviews*, 1862.

10) George Humphry (1820–1896). Professor of human anatomy at Cambridge. Developer of the Cambridge Medical School and the Museum of Anatomy and Surgical Pathology. *A Treatise on the Human Skeleton*, 1858.

11) John Marshall (1818–1891). Holme Professor of clinical surgery at University College, London, and professor of anatomy at the Royal Academy. *On the Development of the Great Veins*, 1850; *The Outlines of Physiology, Human and Comparative*, 1867.

12) James Paget (1814–1899). Surgeon at St. Bartholomew's Hospital, London. *A Descriptive Catalogue of the Anatomical Museum of St. Bartholomew's Hospital*, 1847–1852; *Clinical Lectures and Essays*, 1875.

13) Thomas Watson (1792–1882). Private practice and professor of medicine at King's College, London. *Lectures on the Principles and Practice of Physics*, 1843.

5. STATISTICIANS

In the early nineteenth century statistics developed into a subject with a large following among scientists, politicians, government civil servants, and social reformers. As early as 1833 a section on statistics was added to the British Association and a year later the Statistical Society of London was founded. There had been several things which converged to make statistics popular in the early part of the century. One of these things was the claim by the Belgian statistician Adolphe Quetelet that statistics could be the basis of the science of man and would itself be based upon the laws of mathematical probability. However, the basic reasons for the popularity of statistics in England were, first, the need for information on important social and economic problems and, second, the emergence of a new kind of civil servant aware of the power of statistics in his own field.

Galton received replies from at least seven statisticians and, although their scientific credentials were much stronger than those of the typical statistician, these seven adequately reflected the major fields of statistical inquiry: economic statistics, sanitary statistics, educational statistics, social statistics. The most eminent missing name, of course, was that of Francis Galton, himself. Galton, however, was not primarily a statistician; if he had been forced to classify himself in 1873, it would almost certainly have been as a geographer, anthropologist, or biologist. The most eminent man missing from Galton's sample who owed his reputation entirely to his activity in statistics was William Farr, the superintendent of statistics in the office of the registrar general. Farr's failure to answer is all the more puzzling, because Farr, like Galton, was interested in the biological basis of natural ability and because Galton may have asked Farr for aid on constructing his questionnaire.

The seven statisticians who did reply to Galton's questionnaire were the following:

1) Robert Hyde Greg (1795–1875). Economist, antiquarian, and agriculturist. "On the Presence of the Corn Laws and Sliding Scale," 1841. *Improvements in Agriculture,* 1844.
2) William A. Guy (1810–1885). Vital statistician and professor of forensic medicine at the University of London. Articles on sanitary statistics and statistical method.
3) William P. Hatherly (1801–1881). Lord Chanceller. Translator of Bacon's "Novum Organum." Member of many parliamentary commissions.
4) James Heywood (1810–1897). Educational reformer. *Academic Reform and University Representation,* 1864; *The Recommendations of the Oxford University Commissioners,* 1853.
5) Rowland Hill (1795–1879). Educational reformer and inventor of the English penny postage system.
6) William Stanley Jevons (1835–1882). Political economist and philosopher of science. Professor at Owens College, Manchester. *Principles of Science,* 1874; *Primer of Logic,* 1876; "On the

Study of Periodical Commercial Fluctuations," 1862; *Theory of Political Economy,* 1871.
7) William Newmarch (1820–1882). Banker. Completed Tooke's *History of Prices,* vol. V and VI.

6. ENGINEERS

As with the statisticians there is no doubt that as a group Galton's engineers were much more scientifically oriented than must have been the average nineteenth-century representative of the profession.

It may be said that the subjects with which Galton's engineers were involved represent a good cross-current of English technological activities in the first half of the nineteenth century: bridge designing and harbor improving, railway engineering, military engineering, hydraulic engineering, metallurgical engineering, the manufacture of armaments. It can also be said that, in spite of the membership of all of his engineers in the Royal Society, Galton's sample of engineers provides good illustration of the slowness with which science and technology merged in nineteenth-century England. Galton wrote shortly after the Paris Exhibition of 1867 had convinced many Englishmen that England would either have to have new facilities for the technical education of future generations or fall behind the French, and especially the Germans, in developing a scientifically based industry. The men who replied to his questionnaire were outstanding products of a system of technical training based upon apprenticeship and individual initiative. It was a system which was able to produce some extraordinary men in terms of their practical accomplishments, but was not necessarily equally qualified to produce the army of scientifically trained personnel who would be needed in the future.

Altogether there were eight engineers who replied:

1) William Armstrong (1810–1900). Inventor of hydraulic machinery, artillery and electrical devices. Founder of the engineering and ordinance works at Elswick-on-Tyne. President of the British Association, 1863. "On the Application of Water Pressure as a Motive Power," 1838; *The Industrial Resources of the Tyne, Wear, and Tees,"* 1863.
2) John F. Bateman (1810–1889). Civil engineer. Designer and constructor of waterworks and harbors for many cities. Author of: "History and Description of the Manchester Waterworks," 1844; "On the present state of our Knowledge on the Supply of Water to Towns," 1855. President of the Institute of Civil Engineers, 1879.
3) William Fairbairn (1789–1874). Ship and bridge builder. Improver of boilers and investigator on the properties of the earth's crust. *An Account of the Construction of the Britannia and Conway Tubular Bridges,* 1849. President of the British Association, 1861.

4) Douglas Galton (1822–1899). Sanitary engineer. Captain of the Royal Engineers. Secretary to the railway department of the Board of Trade, 1854. President of the British Association, 1895. *The Construction of Hospitals*, 1869; *Ventilating, Warming, and Lighting*, 1884.

5) Fleeming Jenkin (1833–1885). Professor of engineering, University College, London, and Edinburgh. Research in electrical engineering and publications on literature and drama. *Magnetism and Electricity*, 1873.

6) Charles Merrifield (1827–1884). Principal of Royal School of Architecture and Marine Engineering, South Kensington. Examiner of Education Office. Research in elliptic functions. Many papers in *Transactions of the Institute of Naval Architects* and *Philosophical Transactions*.

7) William Pole (1814–1900). Professor of Civil Engineering at University College, London. Railway engineering and expert on the philosophy of music and whist. *Cornish Pumping Engine*, 1844. Numerous papers in *Proceedings of the Institution of Civil Engineers*.

8) William Siemens (1823–1883). Electrical engineer. Partner in the Siemens manufacturing works. President of the British Association, 1882.

III. COMMENTARY

Galton analyzed the results of his questionnaire for the reader of *English Men of Science* by dividing his discussion into four chapters. In Chapter I he considered the "antecedents" of his scientists and their parentage. In Chapter II he took up the "Qualities" of the scientists, dealing with abilities, personality traits, and other variables like health, independence, religious views, and so on. In Chapter III—really the central chapter of the book—he analyzed the question of the "Origin of Taste for Science." And finally, in the concluding Chapter IV he collected together the remarks of his scientists on "Education." In this Guide it will be convenient to follow Galton's own order as nearly as possible in making comments.

1. ANTECEDENTS

In Galton's first chapter he discussed the antecedents of his scientists, referring to their geographical and class background. He did not, however, provide the reader with a crucial piece of information needed for fully interpreting this material: the birth years of the scientists who replied. Because of the social changes occurring in nineteenth-century England, conditions varied widely over intervals of even just one or two decades. Fortunately the distribution of birth dates of the scientists who replied is easily derived once the names themselves are known.

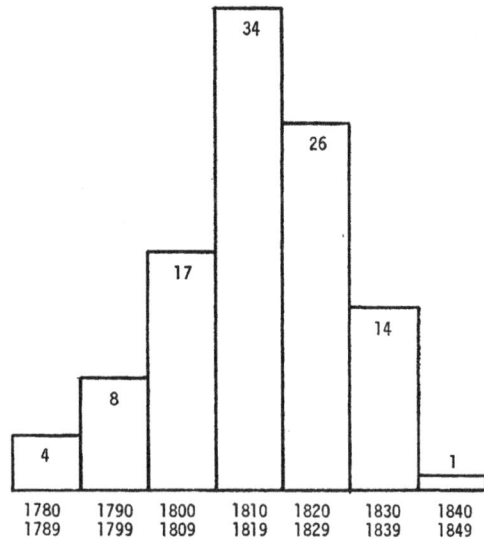

FIG. 4. Number of scientists born by decade.

It turns out that more of Galton's scientists were born in the decade from 1810 to 1820 than in any other decade, but that the years of birth range from the late seventeen-eighties to the early eighteen-forties. The oldest scientist to respond to Galton's questionnaire was Henry Holland, born in 1788, and the youngest was the second generation astronomer, Lord Rosse, who had been born in 1840 and elected to the Royal Society at the young age of twenty-seven, following in the footsteps of his father, also a distinguished astronomer. The actual distribution in terms of decade of birth is shown in figure 4.

Because Galton's scientists were born when they were, Galton's results concerning their generally middle-class and professional background are not unexpected. According to Everett Mendelsohn, "the middle classes (clergy, professions, merchants, civil servants) increased their representation among men of science from 41 percent in the early seventeenth century to 50 percent by the late eighteenth century." [1]

Galton's own results confirm all of this very strikingly. Out of some 96 scientists, according to Galton, there were only 9 "noblemen and private gentlemen." By far the greatest percentage of his scientists came from families involved in business, industry, or the professions. We do not have a key to the replies concerning occupations of fathers, but again it is easy to reconstruct the information once the names are themselves known. Galton was evidently correct in observ-

---

[1] Everett Mendelsohn, "The Emergence of Science as a Profession in Nineteenth Century Europe," in Karl Hill, ed., *The Management of Scientists*, p. 5.

ing the few "farmers" among the parents of his scientists (although perhaps the number should have been three—Fairbairn, Siemens, and Parkes—rather than two).[2] Why such a small number of farmers? Unfortunately Galton's own observation that "mechanicians are usually hardy lads born in the country, [but] biologists are frequently pure townfolk" does little more than accentuate the anomaly, although it is consistent with the fact that both Fairbairn and Siemens were engineers.[3] Whatever the reason, scientists came primarily from urban backgrounds, as the map included by Galton showing the areas of England most productive of scientific talent demonstrates.

2. QUALITIES

In the second chapter of *English Men of Science,* Galton dealt with what he termed the "qualities" of his scientists, meaning by that word character traits, aptitudes, and other things as well. "Some of the qualities," he explained, "are common to all men who succeed in life, others—such as love for science—are more or less special to scientific men."[4] Among the former he included health, energy, memory, and perseverance, whereas the latter embraced "independence," mechanical aptitude, and the like.

Galton led off with ample quotations demonstrating the abundant energy of many of his scientists. Some of the enthusiasm which he felt about the evidence for this energy was reflected at the time by a conversation at the Philosophical Club with the physician James Paget. Paget, who had himself answered one of the questionnaires, wrote:

At the Philosophical, where we had a pleasant dinner, . . . I sat between Hooker and Francis Galton. Galton tells me some very striking facts come out of his enquiries about scientific men. One especially is their very high average of energy and capacity in mere business; the last thing the "vulgar" would expect of those whom they scoff at as philosopher."[5]

A quick examination of either the text or the "Keys" will confirm Galton's impression that on the whole his scientists were extremely energetic in both "body" and "mind." Many illustrated their energy with their dual career in business and science. Many seem to have found it possible to do creative scientific work late at night and with very little sleep. One of the few who found himself deficient in energy was the political economist and philosopher of science, William Stanley Jevons, who wrote, "amount of brain work never remarkable." Galton evidently found Jevons's self-assessment difficult to believe, however, and appended to the quotation from Jevons the corrective that the

"actual performance of this correspondent is considerable, and of a very high order."

The quotations about energy are followed in *English Men of Science* by a paragraph in which Galton argues that energy and head size are inversely related. At first this paragraph will probably seem curious to the modern reader, as it did even to Karl Pearson when Pearson wrote his biography of Galton.[6] Yet, in terms of Galton's own aims and intellectual development the paragraph is very revealing: it hints at the idea of correlation. In the published pages the hint is not very strong (although the word "correlated" is actually used), but in Galton's notebook on "Special Peculiarities," which contains some of the working data, the anticipations of Galton's future work are much more explicit. Particularly significant is a two-dimensional diagram (fig. 5) at the end of the notebook in which Galton compares the distribution of head size (measured by inches round the inside of the hat) with the distribution of stature. For one thing this diagram is consistent with Galton's remark in *English Men of Science* that small-headed men are "not necessarily short," and thus justifies Galton in overlooking the relationship between head circumference and other bodily measurements when discussing the relationship of head size and energy. More importantly, however, the diagram may well be Galton's very first real correlation diagram; certainly it antedates in this respect the 1875 diagram concerning the sizes of peas which has generally been accepted as Galton's first two-dimensional representation of the correlation problem.[7]

---

[2] Francis Galton, *English Men of Science,* p. 22.

[3] *Ibid.,* p. 19.

[4] *Ibid.,* p. 74.

[5] Stephen Paget, *Memoirs and Letters of Sir James Paget* (London, 1901), p. 268.

[6] Pearson writes (*Life, Letters, and Labours of Francis Galton* 2: p. 150), "A modern statistician would not be quite happy in asserting without further investigation that the owners of large heads were less energetic," but he then calculates that the odds are 50 to 1 in favor of an inverse relationship between energy and head size based upon Galton's data. For the record, Galton's raw material is as follows in the notebook on "Special Peculiarities": head size 24″ and above—(Lefevre), Gassiot, Grove, Hirst, Hatherly, Osborn*, Sharpey, Stokes*; head size 22″ and under—Haughton*, Rosse, Sabine, Richards*, F. Galton, Hooker*, Lubbock*, Parkes*, Cobbold*, Bastian*, J. F. Bateman*, D. Galton*, W. Smyth*. Beside each of the names starred Galton has placed an "E" symboling "energy." Beside his own name he has placed a broken "E," symbolizing, one may suppose, either moderate energy or an unwillingness to include himself in the data. Galton, with his own smaller head, often expressed an admiration for the large heads represented at the meetings of the Royal Society; and he felt that energy had to be relatively more important in the success of the smaller-head individuals than in those with large heads. One may note that Galton somewhat overstates his own data in *English Men of Science* when saying that only "one" scientist with head size 24″ or over had great energy, since his data included two such scientists. Pearson's statistics are correspondingly affected.

[7] Usually considered to be the first correlation diagram was that made by Galton for comparing the sizes of seeds of parent and offspring generations of sweet peas—see Pearson, *Life, Letters and Labours of Francis Galton* 2: pp. 392–393; V. L. Hilts, "Statistics and Social Science," in Giere and Westfall, *Foundations of Scientific Method,* p. 40; Helen M. Walker, *Studies in the History of Statistical Method* (Baltimore, 1929),

Galton followed his discussion of energy with a discussion of health. Although Galton quotes sparingly about the health of his scientists, the "Keys" to "health" are fairly complete, and seemingly justify Galton in his conclusion that most of his scientists enjoyed fairly good health. But there was room for misjudgment. It was quite possible for a scientist, as for any other individual, to feel himself very healthy one day and to drop dead the next; indeed, just such a case is told by Galton of the physician Henry Holland in the text of *English Men of Science*.[8] Therefore, we should not be surprised to find disparities such as that represented by the astronomer Richard Carrington, who replied that his health was "delicate till manhood, now excellent," even though we learn from the *Dictionary of National Biography* that some ten years earlier Carrington had suffered a breakdown from which he never recovered. In obvious good spirits, James Clerk Maxwell wrote, "Often laid up before aet. 19, never since; never had a headache." But neither Galton nor Maxwell himself could have foreseen the possibly hereditary tendency in what Maxwell wrote of his mother: "Excellent

health till aet. 48. She died the year after." Although Maxwell's own health and exuberance were excellent until 1879, Maxwell became ill and died in November of that year at the age of forty-eight. At the time of his death, the physician Paget wrote: "It was the disease of which his mother died at the same age."[9]

Except for the brief excursion into the statistical relationship between energy and head size, most of the material concerning "qualities" in *English Men of Science* is anecdotal in character. Thus, for instance, there is a wealth of anecdotes concerning memory, but no statistics, and the reader will willingly agree with Galton's statement that "memory is very variable in power and character."[10] There is considerable indication, however, that Galton had not originally intended to leave his analysis with so few quantitative conclusions; what frustrated him was his failure to get replies concerning the degree to which a quality was possessed. Had it been possible for Galton to have had ratings indicating whether a quality was possessed to degree F (233 in a million) or G (14 in a million) or whatever, he would immediately have drawn up the error curve representing the distribution of the quality. But Galton did not care to push his analysis beyond what the data would warrant, and so only in the unpublished matter in the notebook on "Special Peculiarities" is there evidence for the direction in which he would have liked to have gone.

One statistical excursion reflected in the notebook on "Special Peculiarities" was an attempt to discover the relative importance of the different qualities. For this purpose Galton compared the answers which he received concerning special talents ("mechanism, practical business habits, music, mathematics") and other strongly marked peculiarities ("impulsiveness, steadiness, strong feelings and partisanship, social affections, religious bias of thought, love of the new and marvelous, curiosity about facts, love of pursuit, constructiveness of imagination, foresight, public spirit, disinterestedness"). In descending order of frequency he obtained the following number of cases in which the peculiarity was especially marked:

1. Practical business habits 39
2. Steadiness 35
3. Curiosity about facts 26
4. Mechanics 26
5. Music 18
6. Religious bias 17
7. Disinterestedness 16
8. Social affections 14
9. Love of pursuit 14

pp. 103–104; Ruth S. Cowan, "Francis Galton's Statistical Ideas: The Influence of Eugenics," *Isis* **80** (1972): p. 520. Cowan notes that on an unpublished diagram for his paper on sweetpea seeds Galton "marked the body of the graph wherever parental size and mean filial size intersected" and that thus "the first regression line had been drawn." In fact Galton does exactly the same thing in comparing the distribution of head size and stature at least two years earlier, as is shown in the accompanying diagram from the notebook on "Special Peculiarities." Perhaps, however, one should not put too much stress on this detail. It was during the early and mid-1870's that Galton was learning to think of heredity as having to do with "statistical units"—by which he did not mean simply that biological measurements are distributed according to the law of error, but that inheritance is a matter of one generation passing on a certain fraction of its own deviation from the population average (measured in "statistical units" or number of "probable errors") to the offspring generation. Because this fraction later turned out to be a measure of correlation between the two generations, the most important aspect of Galton's groping towards the correlation concept was his learning to compare two related populations in terms of their variance as measured by "statistical units" "or probable errors." In this light Galton's discussion of possible sexual selection of parents with regard to stature (*English Men of Science*, pp. 31–32) is of relevance to his own development. In this discussion he compared the probable error in stature of the fathers (1.7″) and the probable error in the stature of mothers (1.9″) with the probable error of the average of mothers and fathers (2.3″); since the probable error of the average without selection would be $\sqrt{1.7^2 + 1.9^2} = 2.5″$, he concluded that there was no evidence for selection. Had the predicted probable error of the average without selection not been close to the actual probable error as observed, Galton might have been drawn yet more quickly to the concept of correlation. At least the reasoning behind Galton's discussion of sexual selection in *English Men of Science* was very kindred to the reasoning that led Galton later to write the equation $c^2 = (rc)^2 + v^2$ for the variation (c) of an offspring generation which could be explained by the inheritance of a fraction (r) of the parental variation (c) plus some non-inherited variation (v).

8 Francis Galton, *English Men of Science*, pp. 100–101.

9 Galton's "Key" to health has the figure forty-one for the age of Maxwell's mother at death, but this must be an error. Maxwell's mother died at the age of forty-eight—see Lewis Campbell and William Garnett, *The Life of James Clerk Maxwell* (new ed. abridg. and rev., London, 1884), p. 11.

10 Francis Galton, *English Men of Science*, p. 107.

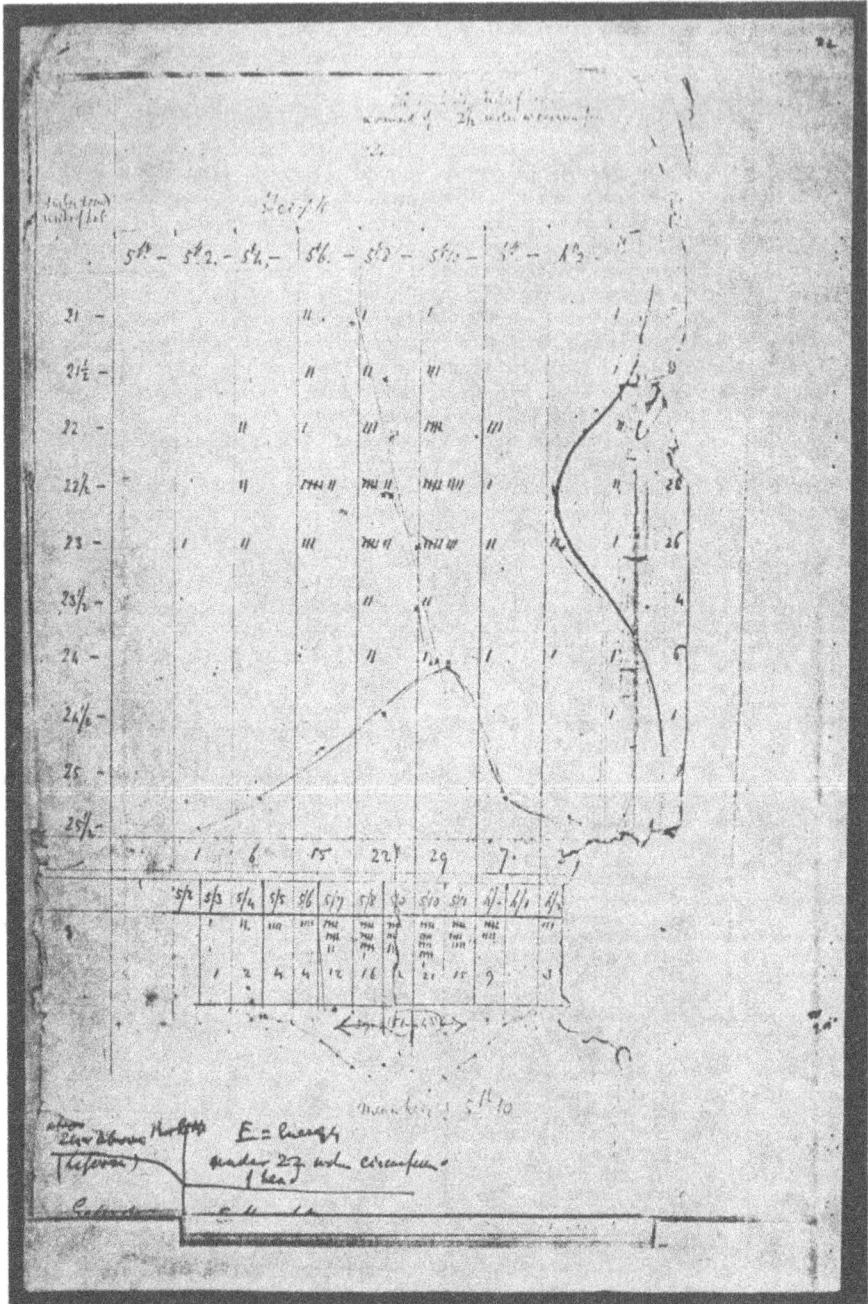

FIG. 5. Head size and height: Galton's first correlation diagram. From notebook on "Special Peculiarities."

An attempt was also made by Galton to determine which characteristics were "acquired or inherited" from the mothers and which from the fathers, as is shown on a chart (fig. 6) in the notebook on "Special Peculiarities." On the top of this chart Galton has listed seventeen qualities (those above plus "partisanship") with three columns under each quality, referring respectively to possession of that quality by Galton's correspondent, by his father, and by his mother. On the side Galton has classified the scientists by scientific field (i.e., sections of the British Association) and noted the number of scientists from each field reporting each quality. At the bottom of the chart are three lines totaling the number of cases "acquired or inherited from both sides," "from father only," and "from mother only." As a result Galton concluded that there were ten cases in which a quality was derived from both parents, some fifty-five cases in which a quality was derived from the father, and some thirty-three cases in which it was derived from the mother; and he noted that it was derived "more than twice as much from the father as from the mother." Since there were some 267 cases of particular qualities being marked among the scientists, and eighty-eight of these were supposedly derived from at least one of the parents, he also noted that about one case in three seemed to be acquired or inherited from parents. Although none of these calculations are specifically mentioned by Galton in his chapter on "Qualities," they seem to be implicitly referred to briefly in the discussion of "Origin of Taste for Science." In that chapter Galton came to the conclusion that about one scientist in four seemingly inherited his taste for science. Compared to the calculation of one scientist in three deriving one of the seventeen qualities from a parent, this would justify Galton in his remark that "instinctive tastes for science are, generally speaking, not so strongly hereditary as the more elementary qualities of the body and mind." [11] The same statistics also seem to be alluded to in the chapter on "Origin of Taste for Science" in Galton's discussion of the lesser influence of the mother. There Galton writes that "in the very numerous answers which have reference to parental influence, . . . the father is quoted three times as often as . . . the mother." [12]

Undoubtedly Galton was wise in not claiming too much for his statistical material and publishing mainly the anecdotal evidence, for the statistics were weak, though suggestive. The anecdotal material also has its weaknesses, but still has considerable value—especially once it is possible to connect the individual and the anecdote. In a few cases—such as those of the botanist Ball, the statistician Hill, the geographer Evans, the physicist Maxwell, and the philosopher Spencer—the anecdotes are very complete and interesting indeed. Why should Maxwell have always heard music, and what tunes did he sing to himself all the time? Was Rowland Hill's insight into the postal system (and invention of the penny postage) at all related to his ability to do cube roots mentally? Was there any relationship between Spencer's castle-building activities as a boy and his later philosophy—or is the latter accountable for by his "unusual faculty of seeing kinship between things not apparently related?" And how was Piazzi Smyth's statement that he had a "preference for whatever is not the fashion, not popular, not rich, not very able to help itself" related to his interest in pyramidology and subsequent resignation from the Royal Society? Is it true that R. H. Scott was totally lacking in "originality," as he claimed—and as the paucity of his work might seem to bear out? The more we know about any particular scientist the more interesting becomes his own self-assessment.

Much of the significance of Galton's material will only be apparent to future biographers of his scientists. Some questions of more general interest are illuminated, however, by pairing scientist and anecdote. In particular, it is interesting to examine in more detail the replies concerning business habits and religious bias.

Considering the middle-class background of his scientists, it is probably not surprising to find business aptitude among the leading qualities reported by his scientists, though it may perhaps be surprising to find that it led all other qualities in frequency of mention. By practical business habits, Galton and his scientists seem to have meant neatness, orderliness, and a general ability to keep things under control—as well as an ability to make money. Galton's statistics concerning the prevalence of business habits were undoubtedly confirmed in his own mind by the case of Charles Darwin, who said that business ability was his only specially marked peculiarity; contrastingly, only James Clerk Maxwell and Justice Grove admitted to having distinctly "low" business abilities. In *English Men of Science,* Galton amplifies his discussion of business ability by noting that some seventeen scientists were also "active heads of great commercial undertakings." [13] Actually, this was a bit misleading, since the seventeen that he had in mind included eight physicians with extensive practices as well as those heading "commercial undertakings" in the strict sense. The eight physicians

[11] *Ibid.,* p. 196.
[12] *Ibid.,* p. 207.
[13] *Ibid.,* p. 105.

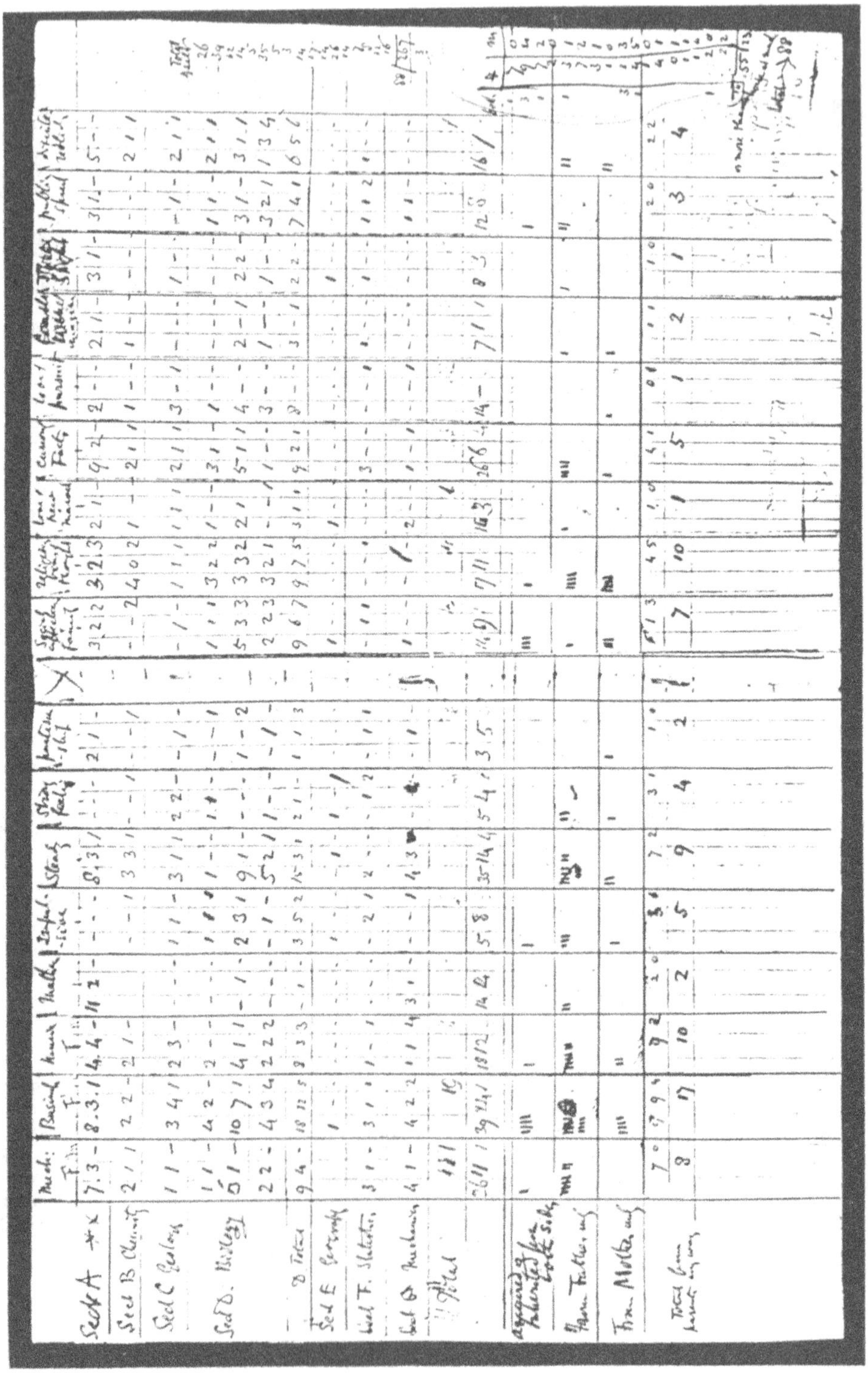

Fig. 6. Qualities of scientists and parents. From notebook on "Special Peculiarities."

were: Alderson, Bowman, Farre, Fergusson, Fox, Humphry, Paget, and Watson. The nine others were De La Rue, Lawes, J. Evans, Osborn, Armstrong, Fairbairn, Gassiot, Spottiswoode and Siemens. Of the latter Armstrong, Fairbairn, and Siemens were engineers with their own firms. Spottiswoode was a printer; J. Evans was a paper manufacturer; and De La Rue was manager of a stationery company. Lawes was a pioneer in the artificial fertilizer industry on his experimental farm at Rothampsted. Gassiot, who seems otherwise not to have been quoted as responding to the questionnaire was a wine merchant.

In point of fact, Galton could easily have expanded his list of seventeen entrepreneurs by including such individuals as Lubbock and Newmarch (bankers), Carrington, Lassell, and Prestwich (brewers), Forbes (who owned his own mining company), Stenhouse (who operated a chemical consulting firm), J. F. Bateman and Miers (engineering firms), and possibly others as well. If these names were added to Galton's own list, the total would be well over one-quarter of the entire number of his correspondents—certainly an impressive number. Such an impressive list might seem to confirm the reality of the Baconian vision in the nineteenth century. However, one must be careful. Of the names just mentioned not all by any means represent a meaningful utilitarian link between science and business. In the case of the engineers, of Lawes, and of Stenhouse, the relationship between science and business is clear; and there would have possibly been more examples of such a relationship had Galton sent his questionnaire to the industrial chemists William H. Perkin and Hugo Müller. But some four of the scientist-entrepreneurs were brewers, whose scientific interests in astronomy (Carrington and Lassell), geology (Prestwich), and physics (Gassiot) seem far removed from their business. Prestwich, in particular, expressed the distance he felt between his daily activities as a wine merchant and his leisure activities as a geologist and archaeologist by writing: "Worked all day at business and found wonderful relief in science at which I worked during many hours of the night." Even De La Rue, who had been trained as a chemist and whose stationery firm employed Müller to do chemical research, spent much of his own time as a scientist occupied with the non-utilitarian field of astronomy (although in this case supported by a specialized knowledge of photography); revealingly, De La Rue wrote that he had "steady business habits but was never 'fond of business." While one cannot discount the possible good effect of business habits on Galton's scientists, it suggests that in some cases science, instead of being closely linked to business, may have been a refreshing antidote to it. Many of Galton's scientists, being the second or third in a line of several generations of businessmen, found it possible to pursue more intellectual activities than had their parents. An extreme case of

this latter attitude was probably represented by Balfour Stewart, who said that he could easily have pursued a business career, but did not because of his "distaste for business." Stewart's father was a tea merchant.

Considering the amount of time that has been spent by sociologists and historians in analyzing the interaction of science and religion, the replies concerning "religious bias" are of great interest. Unfortunately, however, the "Keys" to the section of *English Men of Science* in which Galton gives quotations concerning the possible adverse effect of religion on science have not been located. We do know Darwin's reply on this issue—since he gave up religious views independently at an early age, religion had "no" adverse effect on his science. Because his questionnaire has been preserved, we also know the reply given by Albert Günther to the question—although he considered his theological studies in Germany to have been a waste of time, he still believed that religion, far from being a deterrent to his science, had "acted as a guide." In spite of the failure to locate the "Keys" concerning the effect of religion on science, however, there is a substantial amount of material concerning religious belief: the replies concerning "independence," the replies concerning "religious bias," and Galton's information on religious affiliation. With respect to the latter (which is found in the "Notebook on Head Circumference" and is reported here in the "Index Table"), however, the data are not complete; for instance, some of the scientists who were most vocal in indicating their religious bias—Haughton, Main, Cobbold—do not have their religious affiliation listed.

It is particularly interesting to read the replies concerning religion in light of the thesis advocated by the twentieth-century sociologist Lewis Feuer, who argued —partly on the basis of Galton's material—that scientists are characterized by adherence to a general "hedonist-libertarian ethic," rather than by possession of a religiously inspired "authoritarian super-ego."[14] Feuer notes Galton's statistics concerning the paucity of scientists among the clergy. And he quotes from the replies concerning "independence" to show "the measure of their [scientists'] independence in conflict with society."[15] Although Feuer could not know his identity, it was Balfour Stewart whom he quoted as having written: "Left aet. 12, a school where I received injustice from the master." More relevant yet for Feuer's point would have been Stewart's other remark, as found in the "Keys," that "I entertained at an early age independent views regarding the resurrection and salvation of the heathen, which led to frequent disputes." In fact, although Galton's scientists readily asserted

---

[14] Lewis F. Feuer, *The Scientific Intellectual: The Psychological and Sociological Origins of Modern Science* (New York, 1963), p. 11.
[15] *Ibid.*

their independence, there seems to have been a good deal of ambiguity concerning exactly what was to be meant by the term. In some cases "independence" was associated primarily with religious non-conformity, as by Charles Darwin in his reply that "I gave up common religious beliefs almost independently, on my own reflexion." But at other times something much more general seems to have been meant, as when William Pole, the engineer and musician, wrote: "very independent indeed and will not trust to anybody in authority." Galton himself clearly believed that there was a relationship between independence and religious non-belief, and Feuer's thesis certainly seems to also imply such a relationship. It is interesting, therefore, to compare exactly what was said by the various scientists concerning "independence" with what was said by them concerning "religious bias" to see whether this connection was real, or just how it should be understood. That the connection might be actually somewhat illusory is perhaps hinted by the correspondent—actually the horticulturist James Bateman—who is quoted as having "no more doubt about the plenary inspiration of Scripture than I have about the simplest axiom in mathematics," at the same time as his independence is demonstrated, as Galton noted, by a claim that he "never received or asked a single favor or a single farthing for any thing I ever wrote or did."

Besides Bateman, there are some twelve individuals cited by Galton in the text of *English Men of Science* as having more than just an intellectually inspired "religious bias": Allman, Stokes, Buckton, Ball, J. H. Balfour, Parker, W. C. Williamson, Gladstone, Haughton, Cobbold, and two others who are unidentified. These were evidently the most ardent believers among his group. Only four of these ten names are cited specifically regarding independence, but of these four all claim to be strongly independent, and none of the four mentions a lack of independence. Perhaps the most striking case is that of Cobbold, since Cobbold was a "tract-distributor" considerably involved in religious activities: he replied, when asked by Galton, that his independence was "most decidedly marked." Similarly Haughton, who had been baptised a Quaker but joined the Church of England, replied concerning independence that he had "strong judgments, but cared little for politics." Likewise, W. C. Williamson substantiated his claim to independence by noting that he "Left. Ch. Engl. for Wesleyanism." Finally, John Ball, who was possibly one of only two Catholics answering the questionnaire (George Mivart being the other), wrote that he had independence "but a great distaste to offend the susceptibilities of others."

It therefore appears that for Galton's scientists a self-perceived "independence" and religious belief are not necessarily antithetical; in some cases—as one should certainly have expected—religious bias is bolstered and made possible by an attitude of independence. To some extent, therefore, the thrust of Feuer's analysis seems misguided. Also casting doubt on Feuer's description of the scientific personality (in so far as it applies to Galton's scientists) are the indications that many of the men represented in *English Men of Science* are certainly not examples of individuals "in conflict with society," at least as that phrase has been understood in more recent times. It may be remembered, for instance, that at least a quarter of Galton's scientists were scientist-entrepreneurs: hardly an alienated class. Equally intriguing as an indication of social conformity, rather than conflict, is Galton's mention that a greater number of the parents of his scientists belonged to independent religious sects than did the scientists themselves. Unitarians, Quakers, and Moravians, in particular, were all more prevalent among the parents than among the scientists.[16] Why? By an ironic twist of history it was possible, as has already been mentioned in the case of Samuel Haughton, for an individual to consider himself independent for having left an unorthodox sect, like the Quakers, and having joined the Church of England. However, one may also suspect that some of the sons of nonconformist parents may have found religion less of a living issue than did their parents; and therefore, more as a matter of convenience than conversion, they may have come to terms with the contemporary religious landscape. Perhaps Charles Darwin was not unique in combining religious disbelief with membership—"nominally"—in the Church of England.[17]

[16] Francis Galton, *English Men of Science*, pp. 126–127. Galton writes that seven out of ten scientists were members of the Church of England. He notes that "there is much Quaker, and even some Moravian blood, but there are none who have sent me returns who still profess those creeds." He also notes (p. 127) that the "Unitarian element is stronger" among the parents than among the scientists themselves. However, at least one scientist, W. C. Williamson, left the Church of England in order to become a Methodist.

[17] Another unexpected anomaly concerning religious beliefs came in response to Galton's question, "Has the religious creed taught in your youth had a deterrent effect on the freedom of your researches?" Galton expected the answer to this question to be "Yes." The result, however, was negative. Karl Pearson (*The Life, Letters, and Labours of Francis Galton* 2: p. 151) explained this negative result as follows: "It seems to me that such a question: 'Has the religion taught in your youth had any deterrent effect on the freedom of your researches' can scarcely be accurately answered by the subject himself. If he has cast off the religion of his youth, he may believe that he thinks freely; on the other hand, if he is still a devout believer he would be unlikely to admit that his religious views hampered his scientific researches." Whether or not Pearson's explanation is the true one (and it certainly seems a reasonable one), apparently Galton's scientists did not in fact *perceive* a great conflict between their science and religion; and, therefore, Feuer's description of the Victorian scientist as in "conflict with society" appears overly strong. Incidentally, Pearson mistakenly assumes that the Stokes who answered Galton's questionnaire was the mathematical physicist George Stokes, whereas it was actually the physician William Stokes; Pearson's comments about Stokes' confession of religious bias are therefore irrelevant.

Although Feuer's description of the scientific personality is evidently somewhat overdrawn in application to Galton's scientists, nevertheless, there is some truth to it. Most of Galton's correspondents did consider themselves "independent." And while there is no evidence of "hedonism" of the sort characteristically described by observers of cinema personalities, still many of the scientists did admit to a "love of pursuit" and a "love of truth"—both things also encompassed by Feuer as part of the "hedonist-libertarian ethic." However, one may suggest that each group should be understood primarily within the context of its own time. Galton's correspondents were Victorian scientists, and, perhaps more than anything else, it is the Victorian ethic to which they subscribe. They liked to think of themselves as businesslike. They liked to think of themselves as persevering—especially the engineers among them, who could perhaps all have ascribed to the motto which it is said guided the life of William Armstrong, "Perseverance generally prevails." They cherished high moral ideals, and repeatedly refer to the beneficient examples of their parents on this score. They valued rationality and were willing to work hard in its behalf. They had little patience for the old when it was unreasonable, but not many of them admitted being impulsively drawn to the new and marvelous for its own sake. Like good Victorians and good Englishmen they liked both machines and nature—and saw no conflict between the two.

In the end, Galton's discussion of "Qualities" is refreshing because of its intimacy. We are reminded of the valuable contributions of a Victorian ethic whose negative implications are perhaps too frequently insisted upon by the humanistically minded of a later generation. And we are reminded that each period can find productive stimulus in its own set of values.

3. ORIGINS OF TASTE FOR SCIENCE

Galton's goal was to provide evidence that the most eminent English scientists were possessed of innate, and largely hereditary, abilities. For this reason he gathered the material concerning qualities of both scientists and parents which he presented in Chapter II of *English Men of Science*. But it was really Chapter III, entitled the "Origins of Taste for Science," in which the argument for "innateness" is most forcefully presented. Galton opened the chapter with a question: "What were the motives that first induced the men on my list to occupy themselves with science?" As a result of his analysis, he concluded as follows:

As a rough numerical estimate, it seems that 6 out of every 10 men of science were fitted by nature with a strong taste for it; certainly not 1 person in 10, taken at hap-hazard, possesses such an instinct; therefore I contend that its presence adds fivefold at least, to the chance of scientific success. . . . Nay, further, it appears (though I cannot publish facts in evidence, without violating my rule of avoiding personal allusions) that of the men who

have no taste for science and yet succeed in it, many belong to gifted families, and may therefore be accredited with sufficient general abilities to leave their mark on whatever subject it becomes their business to undertake.[18]

Why should Galton have obtained the results that he did? In the first place, it is worth noting that the replies which Galton analyzed were those given to two separate questions: "Can you trace the origin of your interest in science in general and in your particular branch of it?" and "How far do your scientific tastes appear to have been innate?"

What Galton really wanted to know, and what he assumed he had discovered, was the existence in many of his scientists of a naturally "strong taste" for science. In fact, however, his first question was not really phrased so as to bring this out; it was a purely historical question about the origin of an interest in science. Suppose that a scientist could not trace the origin of his interest in science? When this happened, as it frequently did, it was only natural for the scientist to assume that his interest must be "innate." Thus the real cause of Galton's scientists having so frequently attributed their original interest in science to an "innate taste" seems to be that many of them equated not being able to remember any specific incident that had led them to science to the "innateness" of the scientific impulse itself. Furthermore, since particular influences could rarely be remembered when they occurred at an extremely early age, those scientists whose scientific interests commenced in early youth generally affirmed those interests to be innate.

A few replies will illustrate the equation which was evidently so often made between innateness of taste and early interest in science. The very first scientist quoted by Galton in the chapter on "Origin of Taste for Science" was the astronomer Lassell, who had ground astronomical lenses as a small boy and who wrote accordingly "My tastes are entirely innate; they date from my childhood." What was so succinctly stated by Lassell was reproduced with variations and amplifications by others. Thus the mathematician Arthur Cayley is included by Galton among those citing an innate taste on the strength of Cayley's information that he had "an early taste for arithmetic, and in particular for long division sums." One unidentified geologist neatly illustrates the mental twist which must have taken place in the minds of many scientists by writing: "I believe I may say innate, to a very considerable extent, not remembering that any definite steps were taken to inculcate science." Sir William Armstrong, the inventor, confirmed the trend from the position of the technologist: "If any tastes be innate, mine were; they date from beyond my recollection." Significantly, it is only the physicians and statisticians who do not provide clear examples of the equation of early interests and innateness of scientific tastes, and

---

[18] Francis Galton, *English Men of Science*, pp. 195–196.

these were the very scientists least likely to claim the existence of an innate taste.

If the above analysis of Galton's results is correct, his replies concerning the "Origins of Taste for Science" actually tell us very little one way or the other concerning the inborn scientific aptitudes of his scientists. The replies do, however, raise an interesting additional historical and sociological question. How, in an era when formal scientific education was so frequently lacking, did Galton's scientists manage to become interested in science so early in their youth?

One answer to the above question, which Galton himself considered, was the influence of parental interests. Here the weakness of Galton's questionnaire as an attempt to determine "innateness" of scientific interest is particularly apparent since, curiously enough, an argument in favor of such innateness can follow equally well from diametrically opposed facts concerning parental influence. Thus, if a given scientist found no scientific interest in his immediate family, he was free to believe that his own scientific interests could not have been accounted for by his upbringing and must necessarily, therefore, be innate. Contrariwise, the scientist whose mother or father evidenced an interest in science, could readily believe that his own interest was inherited and therefore innate. An example of the latter reasoning was given by the famous comparative anatomist Richard Owen, who said that his interest in "homology" was innate, while at the same time revealing that it was associated with his "mother's observations in our childhood rambles on the plants and animals we saw."

Galton obviously realized the ambiguity in interpreting the influence of parents upon his scientists, for he wrote in discussing "encouragement at home": "I ascribe many of the cases of encouragement to the existence of an hereditary link; that is to say, the son had inherited scientific tastes, and was encouraged by the parent from whom he had inherited them, and who naturally sympathized with him." [19] Galton also realized that the encouragement at home need not be "direct" (as, for instance, it was in the case Charles Babington, whose father had published an English botany, or William C. Williamson, whose father was a horticulturist) but could take the form of a more pervasive general atmosphere (as in the case of the chemist Roscoe who was granted "permission to carry on little experiments at home in a room set apart for the purpose"). On this basis, Galton calculated that "nearly one-third of the scientific men have expressed themselves indebted to encouragement at home." [20] Reading not only the replies concerning "origin of taste for science" but also those concerning parental interests as recorded in the notebook on "Special Peculiarities," one would suspect that the home influence

must be even greater than estimated by Galton. Certainly there is very little evidence of a generational conflict of interest between Galton's scientists and their parents. In spite of the occasional changes of religion between parent and child, there was expressed a striking similarity between the values of the scientists and their families—quite in contrast to the expressed difference in values between the scientists and many of their schoolmasters.

It thus appears that the parents of Galton's scientists often offered substantial encouragement. When one looks for more specifics, however, the things that stand out are the encouragement of hobbies and collections. With respect to these latter, Galton has already provided more than just a hint of what was happening:

A strong taste bearing remotely on science may prove very helpful. The love of collecting, which is a trifling tendency in itself, common to children, idiots, and magpies, often leads to the study of the things collected, and is of immense use to a man who wishes to study objects that must be collected in large numbers. . . . A taste for drawing works well into engineering and into systematic botany or zoology. A love of adventure and field sports may be an extremely useful element in the character of a man who follows geology or zoology. [21]

More specifically yet, three boyhood hobbies stand out as especially influential on Galton's scientists: mechanics, chemistry, and natural history. In some cases, especially natural history, these hobbies received parental guidance; in other cases, especially chemistry, they were perhaps merely tolerated.

Several scientists mention their interest in mechanics, and this included (as Galton noted) individuals other than the engineers—i.e., Thomas Henry Huxley, who said that he was early attracted to mechanical devices and found natural history against the grain; the physiologist William B. Carpenter, who said that he could have become a "distinguished" engineer (Galton wrote of Carpenter more modestly: "very successful" as an engineer); and Lassell, the astronomer, who referred to his early telescope built as a boy. Carpenter added the information that his early nickname was "Archimedes." Altogether some fifteen scientists are quoted by Galton as having "mechanical attitude" toward things, and at least in several cases it was this attitude which originally spurred on an interest in science.

As important as mechanics in bringing out an early interest in science, it would seem, was chemistry. Thus the mathematician Hirst, the geologists Ansted and Prestwich, and the meteorologist Haughton all seem to have been first caught up in chemistry. In spite of the fact, as Galton shrewdly pointed out, that scientific studies can receive resistance at home because "they deal too much in abstractions on the one hand, and sensible messes and mischief to furniture and clothes on the other," several of his scientists somehow man-

19 *Ibid.*, p. 206.
20 *Ibid.*, p. 205.

21 *Ibid.*, pp. 194–195.

aged to overcome these obstacles.[22] The chemist Roscoe, who has already been mentioned, was one such. Vernon Harcourt, whose father was also interested in chemistry, wrote that "my first taste for chemistry dates from the possession of a chemical box, when I was a little boy." Buckton, another chemist, wrote that "my first notions of chemistry were picked up from books, and I got the nickname 'experimentalizer' at school." An unidentified chemist said that "my father gave me *Henry's Chemistry;* that, and afterward *Turner's Chemistry* were more interesting to me than any books of fiction. . . . I believe at one time I read little else but *Turner's Chemistry* and books of poetry in whatever holiday I had." As opposed to the numerous scientists who mentioned chemistry as a boyhood hobby, there was only one professional chemist who did not become interested in the subject until age eighteen and one other who did nothing at all in science until after he was twenty-three years old.

Probably most significant of all as an influence upon youthful scientific activity—only in part because of the large number of biologists in his sample—were the various boyhood hobbies that could be associated with the countryside: sporting, gardening, animal breeding, the keeping of pets, and simply walking about in the open spaces. Although, as previously noted, there were few farmers' sons among Galton's sample, this was not so negative a factor as it might have been, because Englishmen of all classes placed a high value upon the out-of-doors. We know that both Charles Darwin and Francis Galton, himself, spent a period of their youth during which sportsmanship took precedence over almost all other activities. Among Galton's correspondents, Alfred Newton wrote in reply to the question on the origin of taste for science that "I cannot recollect the time when I was not fond of animals, and knowing all I could learn about them." Noting that his father and brother both had a liking for field-sports, Newton added that "from field sports to field natural history is but a step." Although not making such an explicit claim for field sports as had Newton, the horticulturist James Bateman also characterized himself as "extremely fond of and adept at all field sports." Interestingly enough, even James Clerk Maxwell enters the picture here, for in citing the role played by his father in his early education, Maxwell mentioned his father's love of animals. Much more so than in the esoteric areas of mathematics or physics, or even chemistry, Galton's scientists often received direct encouragement of their own interest in natural history from their parents, many of whom shared a similar interest, even though at the hobby rather than the scientific level. An interesting confirmation of this is the fact that almost two-thirds of those who specifically mentioned their indebtedness to home teaching were either biologists or, nonetheless, like Maxwell

and Jevons, cited their parents' interest in natural history.

4. EDUCATION

Although most of the scientists considered themselves to be "studious" by temperament, almost all had great criticism to make of their formal exposure to schooling. Considering how many of them appear to have become acquainted with scientific pursuits at an early age, we cannot be surprised that they should have keenly felt the lack of such topics in the educational curriculum.

Unfortunately neither Galton nor his scientists distinguished very clearly between comments meant to apply to early schooling and those meant to apply to later education. With respect to early education a great deal of diversity is represented, since the scientists were young in the days before compulsory school education. Some seventeen of them mention teaching at "home" as an important part of their upbringing, with several scientists specifying who in particular was responsible for the teaching at home—sometimes more than one person. In four cases (Bentham, Egerton, Hooker, and Newton) stimulus was provided by tutors at home, whereas in four other cases (Babington, Cobbold, Gray, and Maxwell) the father is cited. Owen noted the influence of an uncle; Armstrong mentioned the teaching of both "mother and sister"; and Newton mentioned "mother, sister, and tutor."

The diversity of educational experiences must be taken into account in assessing the "merits" and "demerits" of schooling mentioned by the scientists. In addition to the fact that some scientists had home education, a few cases of apprenticeship were also represented, as for instance by the physicist Hirst (apprenticed to a civil engineer), the geologist Jones (apprenticed to a surgeon), and the physician Paget (apprenticed to a surgeon). Some scientists did not attend school until they were in their teens, whereas others attended only until that time and had no formal tuition thereafter. Even when the diversity of early educational experiences is taken into account, however, one is struck, as was Galton, with the importance of self-teaching among his scientists. Whereas many of Galton's scientists lacked an opportunity for formal education along the lines they desired, there is no doubt that many potential scientists of a later generation lack the opportunities for self-teaching possessed by many of those in *English Men of Science.* There is much wisdom in Galton's observation that "Much teaching fills a youth with knowledge, but tends prematurely to satiate his appetite for more." [23]

With respect to higher education, the overall situation, though bad, may not have been quite so dismal as appears from Galton's own analysis. Galton himself wrote that "one third of those who sent replies

[22] *Ibid.,* p. 206.

[23] *Ibid.,* p. 257.

have been educated at Oxford or Cambridge, one third at Scotch, Irish, or London universities, and the remaining third at no universities at all"; but he neglected to mention that about fifteen per cent had also received a continental education.[24] There is also need for qualifying Galton's own results because of his failure to get a better return from the theoretical physicists and mathematicians—the very scientists who would have been most likely to be indebted to formal university training.

Because the degree to which the scientists received higher education differed greatly from field to field, as did the assessment which they made of that education, it will be useful to discuss the replies concerning education briefly in terms of scientific fields.

a. MATHEMATICIANS AND PHYSICISTS

It was with respect to the astronomers, mathematicians, and physicists that Galton's replies concerning education were most compromised by failures to reply to the questionnaire.

In the nineteenth century the most important center for English mathematical education was Cambridge, where upon graduation the best mathematicians finished as "wranglers"—the top two being respectively the "senior" wrangler and "second" wrangler. For whatever reason, the highest wranglers are noticeably absent from Galton's list. Living senior wranglers who could have been represented, but who were not, were: the mathematicians Stokes, Sylvester, and Routh; the astronomers Airy and Adams; and the physicist Tait. Missing among the second wranglers is Lord Kelvin. Galton did in fact receive a reply from Arthur Cayley, who had been a senior wrangler (and, incidentally, one of Galton's own Cambridge tutors), but Cayley evidently did not make any substantial remarks concerning his education. Archibald Smith, who had been a senior wrangler, is represented also, but only by a posthumous entry. Thus James Clerk Maxwell, who had been a second wrangler, is left essentially alone as a representative of the highest Cambridge wranglers. Even here, however, there is a lacuna, for Maxwell's remarks concerning his education are confined to his early experiences, and nothing is said about his time at Cambridge. Unlike many of Galton's lesser scientists, who had very great criticisms to make of their early educational opportunities, Maxwell's replies were filled only with untempered enthusiasm and exuberance; more a measure of the man, one suspects, than of the situation.

Although Galton missed getting replies from two astronomers who had been senior wranglers, he did receive a reply from one astronomer who had been a sixth wrangler and one who had been a thirty-sixth wrangler. The sixth wrangler was Robert Main, who

²⁴ Ibid., p. 236.

seems not to have made any striking remarks about his Cambridge education, but who did indicate that between the ages of "13–16 and a half" he was taught by "an excellent teacher" to "reason, use my own mind and depend upon myself." The thirty-sixth wrangler was Richard Carrington, who did have some remarks to make about his higher education. Carrington wrote that he was "put back by too much grounding at Cambridge" and had not enough liberty. He also indicated that his education contained "only one good thing, that was object lessons, though badly given and only a short time." It is perhaps worth observing, however, that Carrington's criticisms of Cambridge are not what we would expect from the biographical sketch in the *Dictionary of National Biography;* there we learn that Carrington had intended a clerical career when the astronomy lectures of Professor Challis at Cambridge directed him into science.

A mathematician who is quoted extensively in *English Men of Science* is Thomas Hirst, whose career, however, was quite unusual. Unlike most English mathematicians of the time, Hirst attended neither Oxford nor Cambridge, but was educated almost entirely on the continent. In 1846 Hirst became a surveyor's apprentice at Halifax, and it so happened that the principal assistant in the same office was the aspiring physicist John Tyndall. When Tyndall went abroad to Marburg to study chemistry, Hirst followed, receiving his Ph.D. there in 1852. Hirst then went on to study at Göttingen, Berlin, and Paris. In Paris, Hirst drew inspiration from the famous French geometer Michael Charles. All in all it was an education more typical of contemporary English chemists than of mathematicians, and Hirst seemed more than satisfied with what he had received.

Among the most interesting educational careers recorded from among the physical scientists are those of the geographers and meteorologists. Although the number of geographers and meteorologists was small, their stories occupy a larger space than might have been anticipated—perhaps because they had some of the most interesting stories to tell about their education in particular and their careers in general. Through opportunities for exploration, the necessity of learning navigation, and the current scientific interest in terrestrial magnetism (stimulated by the researches of Gauss and Humboldt on the continent), the lad who entered the navy at the age of fourteen could possibly become a full-fledged geographer and a member of the Royal Society by his thirties or forties. As would be expected, however, this era of casual but practical education among the geographers was already drawing to a close. In 1873 Edward Sabine (a geographer as well as an astronomer) was eighty-five; George Back was seventy-seven; Frederick J. Evans was fifty-eight; Richard Strachey was fifty-six; and Sherard Osborn was Galton's own age—fifty-one.

The geographer on whom Galton's book sheds the most biographical material is the hydrographer Captain Frederick Evans, whose educational experiences are worth remembering as an example of what could lead a man into science before the middle of the century. Evans entered the navy in 1828 and between 1841 and 1846 as a master of the *Fly* was engaged in surveying the Coral Sea, the Torres Straits, and the Australian Great Barrier Reef. Then in 1855 Evans became superintendent of the compass department of the navy. Together with Archibald Smith, Evans wrote the *Admiralty Manual for Deviations of the Compass,* which first appeared in 1862.

Not all of Galton's geographers were so enthusiastic about their navy education as was Captain Evans. Sherard Osborn, the Arctic explorer, regretted the few opportunities afforded him for study. Osborn was in school for only five years, between the ages of eight and thirteen, and was "taught to read, write, and speak badly." On the other hand, George Back managed in a unique way to pick up what he considered to be an adequate education while in the navy. After having entered the navy at the age of "11¾," Back learned in two and a half years enough to pass the seamanship examination ordinarily taken after ten years. He was then captured by the French during the Napoleonic wars and as a French prisoner had a brief period of formal schooling in languages and mathematics—which accounts for his otherwise puzzling comment that "the years spent as a prisoner were favorable to habits of observation."

b. CHEMISTS

Probably more than any other science except mathematics, chemistry was beginning to emerge as a recognizable university study at the time Galton's scientists were pursuing their education.

Within terms of English educational opportunities, Galton's sample of chemists does justice to the importance of chemical tuition given at the University of London and the University of Glasgow, two of the most important centers at the time in the British Isles. However, his failure to have a yet better sample did mean that only two former students of the Royal College of Chemistry are represented. The Royal College had been instituted in 1845 under the direction of Augustus Wilhelm von Hoffman, a former student of the German chemist Justus Liebig, and was the most evident contemporary attempt to transplant the German system into England. Hoffman, himself, had resigned his position and returned to Germany some ten years before Galton's questionnaire, but among Hoffman's distinguished English students were such scientists as William Crookes, William Henry Perkins, Warren De La Rue, and George Buckton; of these only Buckton and De La Rue had a chance to reply, since the questionnaire was never sent to either Crookes or Perkins.

Galton's sample of chemists confirms what we know from other sources; namely, that the young man really intent upon a serious career in chemistry was likely to find a sojourn to European laboratories a necessity. In fact, the chemist most quoted in Galton's material, Lyon Playfair, found the greatest demerit in his education to stem from the need for just such educational wandering. Having attended some five different universities (three Scottish, London, and Giessen), Playfair complained of the "too great changes of system." Significantly enough, Playfair was later to become a leading spokesman for the improvement of English technical education.

In the chapter on "Origin of Taste for Science" Galton singles out Playfair alone as a chemist mentioning the "influence and encouragement of masters, tutors, and professors." However, one must understand this remark as pertaining only to the initial boyhood interest in chemistry, and not as a comment upon the entire career of the chemists. It is not surprising that only a few chemists would mention teachers in discussing their original early interests, for we have seen that Galton's material supports the view that chemistry was often first indulged in simply as a youthful hobby. But there is ample evidence, even among Galton's material, for the "influence and encouragement of masters, tutors, and professors" when one examines the replies concerning education. Buckton, for example, mentions his time with Hoffman. Roscoe, who had attended the University of London and Heidelberg, wrote that he was indebted to "Prof. Williamson [at London] and to Bunsen [at Heidelberg]." Harcourt, without specifying the influence of any particular individual, noted that he had benefited at Oxford from a "college laboratory at Balliol, then almost untenanted." Stenhouse said that he "he learned natural history under James D. Forbes, who was a very good teacher." Andrews, a physical chemist whose early training had been in chemistry, wrote that "my tastes were partly natural, partly encouraged by an eminent friend"—probably the chemist Thomas Thomson.

c. GEOLOGISTS

On the whole Galton's geologists were among the most dissatisfied with their education of any group in Galton's sample. Ansted, who was a wrangler at Cambridge, did not mention the merits or demerits of his Cambridge education but did claim that he had "been self-taught at home in a desultory way" and that his education had an "absence of necessary control." Smyth, who had also gone to Cambridge, said that he was "almost entirely self-taught as regards natural science" and that "I could now wish I had gone through at the University a good course of chemistry and physics as a preparation for the other branches but main obstacle was lack of time." Prestwich, who had received a second in chemistry at University College, London, regretted the "neglect of many subjects for the attainement of one or two" and "not pushing mathematics to a useful end." Jones, who did not attend a university but was apprenticed to a surgeon, was one of the few who had any good to say for the classics, writing that the "Study of Latin and Greek seemed to give a command over words and phrases." But Jones also indicated that he "learned little at school." Forbes, who had attended Edinburgh, wrote that "a little elementary Physics & Chemistry taught me aet. 7-13 at one school was most conducive to habits of observation." But only Evans, whose father was headmaster of a school, seemed genuinely well pleased with his education, citing its variety of subjects and attention to details.

d. BIOLOGISTS

Galton distinguished between zoologists, botanists, and physicians in his chapter on innate tastes. A distinction ought also to be made between physicians and other biologists in discussing education. Practically speaking it was almost only in medicine that real opportunities for scientific education in biology existed for many of Galton's scientists.

Several of the biologists attended Oxford or Cambridge, but they seem not to have benefited much from their experience at either place. In this respect Charles Darwin is undoubtedly the classic example. In answer to Galton's question concerning the merits of his education, Darwin wrote "none whatever," and furthermore, that "I consider all I have learned of any value has been self-taught." One should, of course, take care not to generalize too rapidly simply from Darwin's story, and it is somewhat hard to believe that his education had no merits "whatever." Nevertheless, some of the other biologists who had gone to the universities echoed Darwin's sentiments, especially in their emphasis upon the importance of self-teaching. Ball, who had been to Cambridge, although his religion made him ineligible for a fellowship, made no comments about his Cambridge experience in reply to the questionnaire but did indicate with reference to an earlier period that "self-teaching has many advantages for boys of active mind." Sclater, who had received first-class honors at Oxford, indicated that his education was "not conducive to observation." Some of those biologists who had not attended universities also emphasized the importance of self-teaching, as did George Bentham, Gray, and Lubbock.

Alfred Newton was one of the few biologists who did express an indebtedness to his Cambridge education, and his story may be set beside that of Darwin for contrast. Until the age of fifteen, Newton was educated entirely at home, and it was while he was still at home that his taste for natural history first developed. Later, however, he attended Cambridge and received a traveling fellowship from Magdalene College. This fellowship, Newton explained, "was tenable for nine years, and its income was sufficient to keep me during that time without being obliged to enter any profession." Evidently the time was not wasted, because in 1866 Newton was appointed to the first chair of zoology and comparative anatomy at Cambridge.[25]

e. STATISTICIANS

A survey of the statisticians reveals that in general they had the best formal education of any group in *English Men of Science*. The only self-educated statistician was William Newmarch, whose competency in economic statistics developed through his employment in banking. Consistent with the low mathematical development of contemporary English statistics, however, few had been trained in higher mathematics. William Guy, who was among the most theoretical of the statisticians, had been a "low wrangler" at Cambridge. Heywood, who was "senior optime" at Cambridge and had attended Edinburgh, wrote about the "thoroughness" of his education. Lord Hatherly, who had been 24th wrangler at Cambridge, wrote only that he had studied natural history at Geneva under De Candolle.

William Stanley Jevons was among the statisticians who made remarks about his educational career, although these remarks seem to have been limited to his early schooling. From other sources we know that Jevons was far from happy with his university training, in part because he was forced to learn the economic ideas of J. S. Mill at a time when he already saw their limitations. It is, therefore, of interest to see Jevons's approval of his earlier education:

I should be much inclined to think there was an innate tendency, but that the tastes were developed by a good and for the most part suitable education. When at my first school, aged ten and a half to twelve years, the headmaster gave very clear occasional lessons in moral and economical subjects. I can remember vividly to the present day the impression which those lessons made upon me. As I am not aware that the other boys in the class were equally impressed, I think I must have had an innate interest in those subjects; but the lessons probably increased the interest very much.

In spite of his generally "suitable education" for a year and a half, however, Jevons was not entirely happy with his schooling. He wrote—as did so many of Galton's scientists—"Latin and Greek grammar, a blank waste of time"—and indicated as a "demerit" his years aged twelve to fifteen at a private school.

---

[25] The geologist Warington Smyth also benefited from a four-year traveling fellowship from Cambridge, which allowed him to travel in Europe, Egypt, and Asia Minor; but Smyth has left no comments about its effect on his career.

f. ENGINEERS

As might have been expected, few of Galton's engineers benefited from much formal higher education. William Fairbairn, who was the oldest of the engineers and eighty-five years old at the time of Galton's questionnaire, wrote that he attended "small schools aet. 6–14 and then became my own schoolmaster." Fairbairn owed his start to his apprenticeship to a millwright and his later friendship with the famous railway engineer George Stephenson.

The story with respect to some of the other engineers was similar. William Armstrong wrote that he attended a large school but "was rather a dull boy and hated Latin and Greek." Armstrong was "a good deal taught at home by my mother and sister." John F. Bateman attended Moravian schools, which, he wrote, were "too general, no science instruction." William Pole attended schools to the age of sixteen and wrote that he was "self-taught in all but the usual education of the English common schools." Two of the engineers were educated on the Continent. William Siemens indicated that he spent "one year at Göttingen—then had to go to business." Fleeming Jenkin attended the University of Genoa between the ages of sixteen and eighteen, where he indicated that he benefited from "natural philosophy and science well taught." However, even Jenkin said that he had been "self-taught in all higher mathematics, mechanics and professional work."

IV. QUOTATIONS AND IDENTIFICATIONS

The raw material available from Galton's *English Men of Science* and associated manuscript material is presented here in three sections: (1) Quotations, (2) Identifications, and (3) an Index Table.

The various manuscripts relating to *English Men of Science* are the following, all of which are located in the Galton archives, University College Library, London:

1) Annotated membership list of the Royal Society for November, 1872. [Galton (1) = 3iiB]
2) Nine "Keys" used by Galton in preparing *English Men of Science.* [GD 18]
3) Notebook entitled "Circumference of Heads of Scientific Men." [Galton (1) = 42]
4) Manuscript of extracts on the "Mental Peculiarities" of various scientists (referred to as "Extracts on Mental Peculiarities") [Galton (1) = 3iiB]
5) Notebook entitled "Special Peculiarities, sorted and classed. Energy and size of head." [Galton (1) = 3iiB]

The references to the items above are the packet numbers as indexed in Galton archives. The University College Library also has a copy of the questionnaire completed by Albert Günther.

The nine different "Keys" made by Galton are the ones on "Business Habits," "Education," "Energy," "Health," "Independence of Character," "Innate Tastes," "Memory," "Scholarliness," and "Steadiness." They are large sheets of paper on which the names are placed (more or less) according to the nature of the reply, and were used by him in preparing the respective section of *English Men of Science*—although they appear to have been made up before the returns had all come in.

1. QUOTATIONS

A great deal more information is available from Galton's material regarding some scientists than others. In the following the relevant quotations for each scientist from both the manuscripts and the text of *English Men of Science* have been brought together. Since Galton has dealt in successive chapters with "Qualities," "Origin of Taste for Science," and "Education," the same order has been observed in presenting these quotations.

Quotations printed in plain type have been taken from the manuscripts, whereas quotations in italics have been taken from the printed text. The various manuscript sources have been indicated as follows:

A : "Key" to "Business Habits"
B : "Key" to "Education"
C : "Key" to "Energy"
D : "Key" to "Health"
E : "Key" to "Independence"
F : "Key" to "Innate Tastes"
G : "Key" to "Memory"
H : "Key" to "Scholarliness"
I : "Key" to "Steadiness"
J : "Extracts on Mental Peculiarities"
K : Discussion of religious bias in notebook "Special Peculiarities, sorted and classed. Energy and size of head."
L : Darwin Questionnaire
M : Günther Questionnaire

The source of textual quotations can be found by referring to the Index Table and the Identifications.

It will be noted that there is some duplication of replies, since the manuscript quotations and identifications overlap. Since the duplication is not excessive, it has been thought better to let it stand rather than indicate those cases (and there are many) in which the wording of the quotations given in manuscript differs at least slightly from that printed in *English Men of Science*. In most cases it is necessary to have both manuscript and printed quotations in order to get the maximum information.

A few words of explanation are necessary with respect to Galton's notational system for referring to relatives in his manuscripts. Those relevant are:

1) In the "Keys" to "Health," to "Independence," and to "Energy," the notations:
   a) "F." = "father"
   b) "f." = "mother"
2) In the "Extracts on Mental Peculiarities" the notations:
   a) "F." = "father"
   b) "M." = "mother"
   c) "f.F." = "father's family"
   d) "f.M." = "mother's family"
   e) "Y.S.B." = "self, sisters, and brothers"

With respect to education, Galton had asked his scientists whether their education was conducive to health or observation, and it is the replies to these two questions respectively which are given under the headings: "to health" and "to observation."

Finally, it is to be understood that since these quotations are taken either from Galton's notes or his text—not directly from the questionnaires themselves —they really reflect Galton's reporting of the scientists' replies. In many cases it seems obvious that Galton is taking his notes verbatim from the questionnaires, but in other cases he merely summarizes or gives extracts.

For stylistic reasons, the quotes from the printed text of *English Men of Science* have not been taken from the first edition, but from the later edition published in 1890 by D. Appleton & Co., of New York. This latter edition, while introducing no substantive changes, clarifies some punctuation and avoids the repeated use of such abbreviations as "aet." and "&c."

JAMES ALDERSON

1. **Qualities: Health:** "Chest affections when young; strong after. F. Vigorous. f. Delicate & phthisical."[D] **Energy:** "Physical—'Very energetic, athletic.'"[C] **Mental Peculiarities:** "No music. Remarkably free from love of the new and marvellous."[J]

3. **Education:** "No large schools. Cambridge, 6th wrangler."[B]

GEORGE ALLMAN

1. **Qualities: Health:** "Good. F. Good till between 60 & 70. f. Good."[D] **Mental Pecularities:** "Social affections; religious bias; love of the new and marvellous; curiosity about facts. F. Social affections; religious bias; disinterestedness. M. Social affections; strong religious bias; disinterestedness."[J]

2. **Origin of Taste for Science:** "Innate love of nature and natural phenomena."[F] *"Innate, as far as a love of nature and of the observation of natural phenomena. I trace the origin of my interest in science to the love of truth and of mental cultivation in my father,*

and his encouragement of this love in his children. I do not think it was largely determined by events after manhood." **Religious bias:** *"Religious bias."*

3. **Education:** "Belfast Acad. Inst., University Dublin & Dublin Medical Schools. Various prizes. Taught much at home. Conducive to observation. Healthful, especially at home. Merits—Home and self-education developed observing faculties. Demerits—School fellows ridiculed studiousness."[B]

THOMAS ANDREWS

1. **Qualities: Health:** "Good. F. Good. f. Good."[D] **Energy:** "Physical—'Active bodily habits long continued but not violent exertion'; Mental—'Continous mental application; was in full practice as physician when lecturing in chemistry and engaged in acid investigations for which medal Royal Society was awarded.'"[C] *"For several years was engaged in full medical practice, and at the same time was a lecturer on . . .[1] and engaged in investigations on . . .[2] for which the Royal medal was awarded by the Royal Society. Father and Mother—Both of active habits."* **Mental Peculiarities:** "Steady and persevering in pursuits once begun; from early life strongly devoted to scientific objects and the same feeling continues. Hope I have acted disinterestedly and have not failed in duty as a public citizen. Organised and delivered for many years, courses of lectures on various branches of science gratuitously to the working classes of Belfast. Have had the character of being able to take clear views of the probable course of future events."[J] **Business:** "Practical habits of business."[A] **Independence:** "Very independent and fearless (refers to papers sent with schedule)."[E] **Scholarliness:** "General science and a taste for general classical literature."[B]

2. **Origin of Taste for Science:** "Partly innate, partly encouraged by an eminent friend."[F] *My tastes were partly natural, partly encouraged by an eminent friend . . . ,[3] who had been honored himself by the friendship of most of the leading men of science in the early part of this century."*

3. **Education:** "Yes—large school, also visited France aet. 15 to study Chemistry. Yes—rather wide. Medicine and chemistry. Prizes. Pursued scientific experiments at an early age. Yes—(Observation). Neither the one nor the other. Favorable to clearness of thought. Careful reading of Aristotle's logic obscurely useful."[B]

DAVID T. ANSTED

1. **Qualities: Health:** "Delicate in early life. One lung seriously affected. After aet. 30 healthy; now extremely healthy. F. Generally perfectly healthy. f. Delicate; died young of bronchial disease."[D] *"Delicate in*

---

[1] Chemistry.
[2] Acids.
[3] Probably the chemist Thomas Thomson (1783–1852), with whom Andrews studied at Glasgow.

early life, one lung seriously affected; mother delicate and phthiscal." **Energy**: "Physical—'Very considerable energy, active habits, power of endurance': Mental—'Great energy in pursuit of one object.'" [C] "*Habitually travel by night without interfering with work of any kind carried on during the day. Active habits and great power of enduring fatigue.*" **Mental Peculiarities**: "Good practical business talent; no taste for a knowledge of music; fair taste for mathematics, cultivated when young. More impulsiveness than steadiness, accompanied by love of pursuit of new subjects and some distinterestedness. F. Good business habits, but not remarkable." [J] **Business**: "Good business talent." [A] **Independence**: "Independent opinions." [E] **Memory**: "Bad for figures, dates and abstract facts." [G] **Scholarliness**: "Not remarkable, but fairly large and varied acquirements." [H]

2. **Origin of Taste for Science**: "Quite a natural taste." [F] "*A natural taste. My interest in science began very early, originating in a love of experiment, at first in chemistry. . . . The ultimate direction of my scientific tastes dates after completion of my regular education.*"

3. **Education**: "No large school. Cambridge—wrangler; Fellow of Jesus. Self-taught at home in a desultory manner. To health—neither way. Absence of necessary control. Want of system." [B] "*Want of system; absence of necessary control.*"

WILLIAM ARMSTRONG

1. **Qualities**: **Energy**: "Physical—'Pretty active and considerably power of endurance'; Mental—'Great energy in pursuit of an object.'" [C] **Mental Peculiarities**: "Very strongly marked talent for mechanism and constructiveness generally, exhibited as a child; little taste for mathematics. Steadiness and perserverance in the pursuit of an object is my most distinctly marked mental percularity. F. An excellent mathematician, but no talent for practical mechanics; strong reasoning power. M. A fine taste for literture and an excellent critic." [J] **Steadiness**: "Steady pursuit of an object is the most strongly marked characteristic." [I] "*Steadiness and perserverance in the pursuit of an object is my most distinctly-marked peculiarity.*"

2. **Origin of Taste for Science**: "If any tastes be innate mine were; they date beyond my recollection." [F] "*If any tastes be innate, mine were; they date from beyond my recollection. They were not determined by events after manhood, but, I think, the reverse; they were discouraged in every way.*"

3. **Education**: "Yes, large schools—was rather a dull boy and hated Latin and Greek. No university. A good deal taught at home by mother and sister. To observation—no returns. To health—no returns. Merits—[blank]. Demerits—Not a good school education, not allowing my mind to follow its natural bias." [B]

CHARLES BABINGTON

1. **Qualities**: **Health**: "Good. F. Good except from results of an accident." [D]

2. **Origin of Taste for Science**: "A natural inclination encouraged by my father." [F] "*To my father's encouragement of a natural inclination.*"

3. **Education**: "Yes, large school. Yes, Cambridge. Not much of anything except a *little* classical learning. Father taught me rudiments of natural science, afterwards taught myself Botany, Entomology and Archaeology. To observation—no returns. To health—no returns. Demerits—Bad early masters; neglect at public school." [B] "*Bad early masters; neglect at public school.*"

GEORGE BACK

1. **Qualities**: **Energy**: "Physical—'Escaped from prison in France 1814. Three arctic land expeditions.'" [C] **Mental Peculiarities**: "From early youth was fond of drawing, which was of great use to me." [J] **Independence**: "Persevering without orders from one Hudson Bay post to another (see particulars); F. a man of decision." [E] **Memory**: "Strong local memory." [G] "*Strong local memory especially of scenery.*"

2. **Education**: "Yes, large school, then to Navy aet. 11¾. In 2½ years had enough knowledge to pass the usual seamanship exam—the ordinary period being 10 years. Was selected from among many volunteers. When prisoner in France attended schools, but from aet. 19 was self-taught. To observation—Imperfect education increased my desire for more. Demerits—Want of being thoroughly grounded. This gave me great trouble but made me think for myself, often an advantage to me." [B] "*The demerit of my education was the want of being thoroughly grounded; this gave me great trouble, but made me think for myself—often an advantage to me.*"

JOHN H. BALFOUR

1. **Qualities**: **Health**: "Generally good. F. Always good till last days of life. f. Good for greater part of life, then paralysis." [D] **Mental Peculiarities**: "Great idea of order and method in business; regular in all my duties; fond of science from my youth. Impulsiveness, strong feelings sometimes leading me too far; much religious bias of thought from early education. At school I was often called 'Fipperty Gibbet.' [4] F. Was fond of science and early interested me in botany during walks in the country; he also was impulsive and strong in his social and political feelings; markedly religious; became a strong supporter of Presbyterianism and a leader in Evangelical circles in Edinburgh." [J] **Religious Bias**: "*Much religious bias of thought from early education.*"

2. **Origin of Taste for Science**: "*The love for botany was instilled into me in very early youth by my*

---

[4] "Fipperty Gibbet" = "Flibberty Gibbet."

father. We lived in the house of . . . [a very eminent geologist] [5]*, in the vicinity of . . . ,*[6] *and I often took walks to those hills and collected plants. I also cultivated plants in our garden. A taste for natural science, especially botany, seems to have been innate. The companionship of . . .*[7] *incited me to prosecute botany with vigor. I was one of his best pupils, and traveled over a great part of . . .*[8] *with him."*

THOMAS G. BALFOUR

1. **Qualities: Health:** "Very delicate as a child from croup. Very healthy since. Only 10 days sickness in 37 years. F. Always good. f. General good health." [D]

JOHN BALL

1. **Qualities: Health:** "Delicate as a boy, subsequently good. F. Excellent until aet. 70. f. Very delicate but strong consititution." [D] **Energy:** "Physical— 'Considerable activity but not much bodily strength, passion for mountaineering (founder of Alpine Club, 1st president of it)'; Mental—'Rather considerable, worked for 16 hours a day during Irish famine.' " [C] **Mental Peculiarities:** "Facility for mental arithemtic from early youth, 8 or 9 years; relations of figures carried on in the mind, as chess without the board; a passion for music but no retentiveness of ear for tune; considerable ability for acquiring languages but mainly, I think through the imitative faculty. I believe a considerable faculty for the quick dispatch of administrative business. A character strongly disposed to look forward hopefully to the future; a perhaps excessive dislike to quarrels and disputes, save *purely* intellectual discussion; much fondness for society of agreeable and sympathetic persons; early religious impressions strong but have on the *dogmatic side* quite disappeared; the belief in a permanent antithesis between good and evil irrespective of utilitarian results has survived, with no keen sense of a dogmatic basis for the belief; strongest aversion is for men who sacrifice public interests for low motives; greatest pleasure in life, enjoyment of natural scenery. F. Slight fondness for music; facility for acquiring languages and catching accent and pronunciation; no facility for arithmetic, geometry or mechanics; slow (over-cautious), in forming a decision except on matters with which he was already familiar. Habitual caution and reserve as to serious subjects; only strong bias shewn, was that of contempt for pretenders and political adventurers. His ideal standard that of the *gentleman* rather than the sage or patriot. Very fond of society and society fond of him. M. Some mechanical facility; absence of facility for music or arithmetic; none for speak-

ing foreign languages though readily reading them. Impulsive; strong partisan, considerable ardour of temperament, not much regulated; strong religious impressions and belief but not much fondness of religious observances. f. F. The main characteristic that of prudence and caution combined with considerable practical capacity. f. M. Celtic temperament; impulsive; exaggerated expression of views founded rather on appearances rather than reality; love of adventure." [J] **Business:** "Quick administrative dispatch." [A] **Independence:** "Yes, but a great distaste to offend the susceptibilities of others." [E] **Memory:** "Very great in some respects, blank in others (? see original)." [G] *"Memory most treacherous except in certain respects. Vivid and generally very accurate as to places and visual images. As to thousands and perhaps tens of thousands of specimens and plants, can remember the exact spot where each was gathered. As to a multitude of facts that should have interested me, my memory is a blank and the original impression revived with difficulty, if at all . . . very retentive and accurate as to the sequence of impressions from early childhood onward.* Father— *Remarkably retentive memory: quoted long passages from classical authors not seen for a very long time previously. Shortly before his death, at seventy-three, recited a long passage from 'Gibbon,' not read for fifty years before.* Mother—*Memory not reliable generally, but clinging strongly to special scenes and events."* **Scholarliness:** "Very inquiring in early life." [H] **Religious Bias:** *"Early religious impressions strong, but have on the* dogmatic side, *quite disappeared. The belief in a permanent antithesis between good and evil, irrespective of utilitarian results, has survived, with no keen sense of the need of a dogmatic basis for the belief."*

2. **Origin of Taste for Science:** "Decidedly innate: I had no external stimulus." [F] *"As far as the word applies in any case, I should say decidedly innate. Excepting such influences as a little encouragement at home, I am unable to trace any external stimulus. At the age of six I was given Joyce's 'Scientific Dialogues,' which I soon mastered, then other books; before the age of eight I commenced making star maps: at twelve to thirteen years of age I made some geological sections with tolerable correctness; and so on. It [then] seemed as if any accident and the love of new vistas were enough to lead me from one branch of science to another."*

3. **Education:** "Yes, large school. A smattering of classics and mathematics. Cambridge. Not being eligible (on religious grounds) for a fellowship pursued a discursive time, was a low wrangler. Till aet. 13 was left to myself—read miscellaneously and voraciously on Phys. Science and Nat. History, and nothing amused except that. Observation—going to Switz. aet. 7 & 11 developed love of scenery. Merits—As compared with ordinary schools, think self-teaching has many advantages for boys of active minds. Demerits—but intelli-

---

[5] James Hutton? Hutton was a near relative, and J. H. Balfour's full name was John Hutton Balfour.
[6] Edinburgh?
[7] Probably the botanist, Professor Graham, to whose chair at Edinburgh Balfour succeeded in 1845.
[8] Scotland?

gent teaching and insisting on accuracy & completeness would have produced a much more efficient man." [B] *"Left to myself, and I pursued a discursive line. As compared with ordinary schools, I think self-teaching has many advantages for boys of active minds; but intelligent teaching, and insisting on accuracy and completeness, would have produced a much more efficient man."*

HENRY C. BASTIAN

1. **Qualities: Health:** "Delicate in early life but gradually getting strong. F. Good. f. Good." [D] **Mental Peculiarities:** "Practical business habits so far as requirements go; no ear whatever for music; cannot distinguish one tune from another. Steadiness of purpose,—love of knowledge and pursuit; foresight; power and love of generalization. Y.B.S. Quickness of perception uniform. f.F. Abilities generally decidedly above the average. f.M. Abilities above the average; independence and decision of character; quick perception and great retentiveness of memory." [J]

JAMES BATEMAN

1. **Qualities: Energy:** "Great energy and power of enduring fatigue." [C] *"Remarkable energy and activity of body, and power of enduring fatigue and going without food. Extremely fond of and an adept at all field sports. Abstemious. Of mind—Vigorous pursuit of scientific experiments and investigations, of investment and management of money, business transactions, etc. Father—Active in field sports; has ridden sixty miles before dinner. Abstemious. Engergetic in mind. Mother—Much energy, as shown by activity and power of enduring fatigue. Great physical courage and presence of mind in danger."* **Mental Peculiarities:** "Always fond of natural history, especially, of orchids; good amateur landscape gardener; not a bad amateur lecturer. . . . As to disinterestedness, I have never received or asked a single favor or a single farthing for anything I ever wrote or did." [J] **Independence:** "Perfectly independent in judgment and party." [E] **Memory:** "Used to be good." [G] **Scholarliness:** "Considerable." [H] **Religious Bias:** *"I have no more doubt about the plenary inspiration of Scripture than I have about the simplest axiom in mathematics."*
2. **Origin of Taste for Science:** "I was always fond of plants." [F] *"Always fond of plants."*
3. **Education:** "Large schools. Oxford, 4th class. To health—fairly so, but too sendentary while at Oxford." [B]

JOHN F. BATEMAN

1. **Qualities: Health:** "Always good. F. Always good. f. Good." [D] **Energy:** "Physical—'Has done his chief brain work between 10 p.m. and 2 a.m., besides all the day labour. Rests perfectly at night in a railway

carriage.'" [C] *"Has done his chief brain-work between ten p.m. and two a.m., besides all the day-labour; rests perfectly during a night railway journey. Father—Great energy, and very active; capable of enduring great fatigue."* **Mental Peculiarities:** "Drawing; not musical. F. Mechanical genius. Musical, fond of horticulture (unfortunately for his career), was a spoiled child in every respect. M. Musical, highly accomplished. Y.S.B. I am the only one who is not musical. f.M. Musical, highly accomplished, strong religious bent." [J] **Memory:** "Great for figures; can get up whole pages for examination before committees." [G] *"Great for figures; can get up pages for examination before committees, and dismiss them from memory afterward. Strong recollection of scenery."*
2. **Origin of Taste for Science:** "Natural tastes which fell in with family tradition through my mother's side and with my pursuits." [F] *"Family tradition derived through my mother's side. My profession fell in with my natural tastes, such as sketching."*
3. **Education:** "Yes, Moravian schools, too general, no science instruction. Self-taught to a great extent. Too desultory but gave wide interests." [B] *"No sound instruction; the education was general and desultory, but it gave wide interest."*

GEORGE BENTHAM

1. **Qualities: Health:** "Always good. F. Always good though over-worked in body at one time and in mind at another. f. Always good." [D] **Energy:** "Physical—'Strong, walked 50 miles a day without fatigue and 5 miles an hour for 3 hours.'" [C] *"Strong when young—walked many a time fifty miles a day without fatigue, and kept up five miles an hour for three or four hours. Father—remarkable energy of body up to the age of thirty, as shown . . . Of Mind—Remarkable energy from early youth to his death (brought on by accident at seventy-three), when he was as actively engaged as ever in preparing for experiments [official and of a very multifarious kind]. Mother—Remarkable energy of mind in assisting her father in the preparation of his lectures, and afterward her husband in his official correspondence and writings. After his death she wrote largely in magazines, and at the age of eighty-five published 'Suggestions for . . . [certain improvements in administration].'"* **Independence:** "Felt deficient when acting as police magistrate." [E] **Memory:** "Retentive of botanical names." [G] *"Retentive of botanical names; rather deficient in other respects, especially as to persons."*
2. **Origin of Taste for Science:** "No." [F] *"Not innate. I trace the origin of my botanical tastes to leisure; to the accidental receipt of De Candolle's 'Flore Française,' while resident in that country; and to encouragement from my mother. They were determined afterward by independence (considering my absence of ambition to rise in the world), and by friendship and*

encouragement from . . . , the four greatest British botanists of the day."

3. **Education:** "Never at any school or university—masters at home when at St. Petersburg, London, and in France; self-taught in special pursuits. Observation—neither way, Health—yes, fairly. Merits—none. Demerits—a public school might have overcome shyness." [B]

WILLIAM BOWMAN

1. **Qualities: Health:** "Good. F. Good. f. Good." [D] **Mental Peculiarities:** "(No returns.) F. Methodical as a man of business; economical of time. Quiet, unambitious and moderate (Very much attached to science, F.G.)." [J]

CHARLES BROOKE

1. **Qualities: Health:** "Very good. F. Good but rather feeble. f. Very good." [D] **Mental Peculiarities:** "Aptitude in contrivance; in adapting means to ends. e.g.—the invention of self-registering magnetic apparatus. F. Considerable taste in art; good water-colour draughtsman." [J]

GEORGE BUCKTON

1. **Qualities: Health:** "Health & constitution good. F. Health & constitution good. f. Early life, health good, but in last years insane." [D] **Mental Peculiarities:** "Mechanism, as shewn by constructing a reflecting telescope with 12 inch aperature [N.B. his special science has no connection with astronomy, Ed.] [9]; fond of music and fine arts; a fair amateur artist in oil colour. Not impulsive; religious views liberal, but strongly nonmaterialistic; rather impatient of new scientific theory unless experimentally supported. F. Considerable business habits, fair musical taste. Steady of purpose. M. Strong religious convictions." [J] **Business:** "Strong business habits." [A] **Mechanics:** "Constructed a reflecting telescope, with 12-inch aperture." **Religious Bias:** "Religious views liberal, but strongly anti-materialistic."

2. **Origin of Taste for Science:** "Perhaps wholly innate." [F] "*Perhaps wholly innate. My first notions of chemistry were picked up from books, and I got the nickname 'experimentalizer' at school. My taste for zoology arose through friendship with. . . . My tastes were largely determined by three years' voluntary work at chemistry, under Dr. . . .*[10]"

3. **Education:** "Partly educated with 4 other pupils with a clergyman, partly at home. Observation—in no way inculcates. Health—no remarks. Merits—such as classical education brings. Demerits—no remarks." [B]

---

[9] Later Buckton turned more intensively to astronomy, and in 1882 "he fell in trying to reach the long focus of a Newtonian telescope, fracturing his leg in two places, and lying for some hours undiscovered." *Dictionary of National Biography.*
[10] Hoffman at the Royal College of Chemistry.

WILLIAM B. CARPENTER

1. **Qualities: Mental Peculiarities:** "Always fond of construction; school nickname 'Archimedes'; If I had followed my own bent, should have probably been a distinguished engineer. Have always been accounted a good 'man of business.' Was not taught music, but was always fond of it, and taught myself the organ after I was 20, so as to play with taste and expression, though I never acquired any power of execution; have always had more pleasure in sacred than secular music, which perhaps shows the predominance of the emotional tendence. F. An extremely good man of business; delighted in order and regularity, and upheld this strongly in his house and school, was fond of constructive arrangements, physical and moral. Was always looked to by the Liberal party in Exeter and Bristol, as one of their most valued leaders in political and philanthropic action. Had a decided feeling and love for music, but never cultivated it. The *sense of duty* was the predominating feature of his character, as it was also of my mother. I should say that my father's intellectual character and my own are rather notable for talent than for genius; we have shewn a good deal of constructive power, but nothing of creative. M. [Strong sense of duty, see above]. Y.B.S. My father combined in a remarkable degree, literary and scientific ability (shewn by his distinguished career as a student at Glasgow, and by his subsequent career) and active philanthropic character. These qualities were *distributed,* as it were among his children [facts are given in evidence]." [J] **Mechanics:** "*Always fond of constructing; school nickname, 'Archimedes.' If I had followed my profession should probably have been [very successful as] an engineer.*" **Religious Bias:** "*I have, perhaps, a religious bias from early habits and associations, rather than from temperament; but I have always had more pleasure in sacred than in secular music, which, perhaps, shows the predominance of the emotional tendency.*"

2. **Origin of Taste for Science:** "*My school nickname was 'Archimedes;' I was always fond of construction. If I had followed my own bent, I should probably have been [successful as] an engineer. My turn for scientific inquiry led me in early life to systematize and generalize the knowledge of others. Latterly I have felt more interest in original investigations.*"

RICHARD CARRINGTON

1. **Qualities: Health:** "Delicate till manhood, now excellent. F. Delicate but pretty good. f. Excellent health." [D] **Energy:** "Mental—'Considerable.'" [C] **Mental Peculiarities:** "I was noted at school as always learning a trade, such as chair-making, boat-building, and printing. I have some capacity for business as shewn by the success I acquired at Brentford, where I succeeded to a concern which never 'paid' from the

beginning, till I took it 40 years after.[11] f.F. My grandfather was successively in seven businesses and succeeded in them all." [J] **Business:** "Business habits." [A] **Memory:** "Very poor." [G] *"I Have a very poor memory; I was once a whole fortnight in recovering the name of . . . , but I got it at last. I consider that all attempts at making me learn poetry, and in particular Latin poetry [at school], were gross mistakes; I was never benefited in the least. Reasoning was my forte, and I could never do anything by rote."* (identification uncertain)

3. **Education:** "Schools small & large—was 1st boy in one of the latter. Cambridge 36th wrangler. Health—generally the reverse but had gymnastics. Merits—only one good thing, that was object lessons though given badly & only a short time. Demerits—not enough liberty, put back by too much grounding at Cambridge." [B] *"Only one good thing; that was object-lessons, though given badly, and only for a short time." "Not enough liberty; put back by too much grounding at Cambridge."*

HENRY J. CARTER

1. **Qualities: Health:** "Throughout life has been moderate, hardly good. F. Died aet. 32 of cancer, previous health unknown. f. Good throughout, now aet. 84." [D] **Mental Peculiarities:** Determination to succeed if possible; perseverance, resourcefulness, foresight. My motto being 'whatever thy hand findeth to do, do it with all thy might,' for 'the night soon cometh in which no man can work.' " [J] **Steadiness:** "Determination to succeed if possible." [I] *"Determination to succeed when possible; my motto being, 'Whatever thy hand findeth to do, do it with all thy might,' for 'the night soon cometh when no man can work.' "*

2. **Origin of Taste for Science:** "I was always fond of objective and experimental knowledge." [F] *"Was always fond of objective and experimental knowledge. I date my first efforts to any consequence from an early intimacy with prof . . . , whose pupil and assistant I was. I had a fondness for science before, but the necessity for accurate and rigid observation then first dawned upon me. Subsequent events were going to . . . [abroad], and appointments in . . . [a foreign country, where I was much detained in-doors, that] compelled me to take to the microscope and study of the lower orders of plants and animals, many of which I could grow in my own room."*

3. **Education:** "Schools, small and large. University College, London. Medical courses, 2nd place medal. aet. 14–16 at home, also 1 year in Paris. Self taught a good deal at all times. Conducive to observation. Ditto health. Merits—forced accuracy in delineation from 14 to 16 at home. Society of my guardian, a literary man of great ability and taste. Demerits—early teaching of Greek and Latin omitted. Idleness at school. Up to 14 I seem to have learnt nothing, then I began to learn quickly at home under the direction of my guardian." [B] *"Forced accuracy of delineation at home, between the ages of fourteen and sixteen."*

ARTHUR CAYLEY

1. **Qualities: Health:** "Good. F. Good. f. Good." [D] **Energy:** "Physical—'Have been a good walker and fond of mountaineering'; Mental—'no returns.' " [C] *"When under twenty, have walked twenty miles before breakfast; when about thirty-two, walked forty-five miles; dined and danced till two in the morning without fatigue. At the age of twenty-six during fourteen days, was only three hours per night in bed, and on two of the nights was up all night preparing for . . . [certain scientific work]. Fond of mountaineering."* **Scholariness:** "No returns [Very great]." [H]

2. **Origin of Taste for Science:** "I had an early taste for arithmetic." [F] *"An early taste for arithmetic, and in particular for long-division sums."*

3. **Education:** "Small & large schools, Kings College, London. Trinity College, Cambridge. Senior wrangler." [B]

THOMAS COBBOLD

1. **Qualities: Health:** "Uniformly good, interrupted only by overwork aggravated by peculiar condition. F. Excellent generally. f. Excellent." [D] **Energy:** "Physical—'Excelled in athletics at school & college. Jumped 18 feet'; Mental 'Almost incapable of fatigue up to aet. 38. Usually engaged in work to 1, 2, or 3 am.' " [C] *"Excelled at school and college in athletic sports, especially in long jumping (eighteen feet). In mind—Almost incapable of fatigue up to the age of thirty-eight. Usually engaged in literary work until long after midnight. Father—Remarkably active habits; a great reader when not engaged in drawing and writing."* **Mental Peculiarities:** "Public speaking, lectures usually with few notes; clear enunciation; drawing, has constructed extensive series of original illustrations of animals and vegetable forms on a large scale. Music; possessed of a very powerful alto voice, in addition to the ordinary chest notes which are tenor and above the average strength; good ear for music; very persevering; not discouraged by defeat; intensely religious and formerly in the Evangelical sense (a tract distributor and promoter of prayer-meetings, bible classes, &c). Excessive distaste to novels and fictions in any shape. F. Impulsive; highly imaginative; very strong religious feeling.[12] M. Good ear for music. Not impulsive,

---

[11] Carrington's father was the proprietor of a brewery at Brentford.

[12] Cobbold's father published several books of sermons and religious pieces as well as a novel. "He was devoted to the church of England, always ready to impress its doctrines on others by example and exhortation." *Dictionary of National Biography.*

sound judgment; not imaginative; no strong religious feelings; distrusts religious persons. f.F. All his brothers rather impulsive, imaginative and talented. 'My numerous relatives though unknown to the scientific world are most of them characterised by greath breadth of thought and rare independence of action. . . . I believe (from good grounds as stated) that these peculiarities were derived from my great grandfather (father's father's father).'" *Independence*: "Most decidedly marked; F. Perfectly fearless." [E] **Memory**: "Strong to aet. 38. Still able to name and recognize from 2000 to 3000 species." [G] *"Memory strong up to the age of thirty-eight; still good and capable of recognizing and naming probably between two and three thousand species of animals and plants, including fossil forms. Father—Remarkable; capable of accurately repeating from memory the substance of speeches delivered at clerical and other meetings."* **Steadiness**: "Very persevering." [I] **Religious Bias**: *"Intensely religious; formerly in the evangelical sense (a tract distributor, promoter of prayer-meetings, Bible classes, etc.) Excessive distaste to novels and fictions in any shape."* **Indifference to Dogma**: *"Not in the slightest degree; but the method of science has taught me not to put any confidence in creeds or dogmatic statements of any kind."*

2. **Origin of Taste for Science**: "Yes. I believe my tastes were inherited from my mother's father." [F] *"I believe I inherited my general taste for scientific pursuits from my grandmother; but my choosing . . .*[13] *for special investigation resulted from a positive fascination which the very obscurity of the subject exerted upon my mind. It was perhaps a mere desire to unravel the marvelous. My scientific tastes were largely promoted by the attractive teaching of [. . . various professors]."*[14]

3. **Education**: "Small & large schools. Charter House, Edinburgh University. Medical and other course. Gold medal for thesis. Largely indebted to my father's instructions for ability as a draughtsman. Decidedly restrictive to observation. School system was depressing to health. Merits—Style—method of teaching at Edinburgh—Also the industrious habits inculcated at Norwich by Crosse,[15] F.R.S., to whom I was articled as a pupil. Demerits—Absence of instruction in the modern languages." [B] *"Absence of instruction in the modern languages."*

FREDERICK CURREY

1. **Qualities: Health**: "Average. F. Fair health, but headaches ind indigestion. f. Weak health; hardly

an invalid but could never take active exercise." [D] **Energy**: "Physical—'Not energetic, consitutionally languid. Strong wish for more power'; Mental—'Yes as far as health persists. Much occupied professionally but work vigorously in botany in leisure hours.'" [C] *"Constitutionally languid, with a strong wish for greater energy and more power of enduring fatigue. In mind—Energetic in body as far as health permits. Much occupied professionally, but, when well, capable of vigorously following up the science of . . . in leisure hours. Father—Energetic in body as far as his health allowed; in mind, very energetic. His brain-work from an early age was very large in amount, and he was vigorous and sanguine about any thing he undertook. Mother—Very languid; incapable of any bodily exertion. Very little energy of mind; too languid to take much interest in any thing beyond her own family."* **Mental Peculiarities**: "Strong curiosity about facts connected with the science of botany." [I]

2. **Origin of Taste for Science**: "No." *"I cannot trace the origin of my interest in any particular branch of science further than that as regards . . . botany, I was thrown into the society of a gentleman who took much interest in it. My scientific tastes originated, as a matter of fact, after leaving . . . [the university.]"* (identification uncertain)

3. **Education**: "Small & large schools, Eton. Scholar Trinity College." [B]

CHARLES DARWIN

1. **Qualities: Health**: "Good when young—bad for last 33 years." [L] **Energy**: "Physical—'Much activity while I had health; power of resisting fatigue; I and one other man were alone able to fetch water for a large party of officers & men utterly prostrated by fatigue'; Mental—'Vigorous & long continued work on one subject as 20 years on "Origin of Species" and 9 on Cirripedia'" [C] *"Energy shown by much activity, and, while I had health, power of resisting fatigue. I and one other man were alone able to fetch water for a large party of officers and men utterly prostrated [other facts given in illustration of undoubted energy]. In mind—Shown by vigorous and long-continued work on the same subject, as twenty years on . . .*[16] *and nine years on . . .*[17] *Father—Great power of endurance, although feeling much fatigue, as after consultations after long journeys; very active; not restless. In mind—Habitually very active, as shown in conversation with a succession of people during the whole day."* **Mental Peculiarities**: "None except for business, as evinced by keeping accounts, being regular in correspondence, and investing money very well; very methodical in my habits. Steadiness; great curiosity about facts and their meaning; some love of the new and marvellous. F. Practical business; made a large fortune and incurred

---

[13] Helminthology.

[14] Cobbold was assistant to Professors Bennet and Goodir at Edinburgh, and was influenced by the lectures of Edward Forbes.

[15] J. G. Crosse, F.R.S., surgeon to the Norfolk and Norwich Hospital.

[16] *Origin of Species.*

[17] *Cirripedia.*

no losses. Strong social affection and great sympathy with the pleasures of others; sceptical as to new things; curious as to facts; great foresight; not much public spirit; great generosity in giving away money and assistance. M. Said to have been very agreeable in conversation." [J] **Business:** "Special to business as evinced by keeping accounts regular in correspondence and investing money very well. Very methodical." [A] *A most eminent biologist wrote as follows, in reply to the inquiry whether he had any special tastes bearing on scientic success, in addition to those for his own line of investigations: "I have no special talent except for business, as evinced by keeping accounts, being regular in correspondence, and investing money very well."* **Independence:** "I think fairly; I gave up common religious beliefs almost independently, on my own reflection; F. Freethinker in religious matters." [E] **Memory:** "Very bad for dates and learning by rote but good for a vague retention of facts." [G] *"Memory very bad for dates and for learning by rote, but [extraordinarily] good in retaining a general or vague recollection of many facts. Father—Wonderful memory for dates; in old age he told a person, reading aloud to him a book only once read in youth, the passages which were coming; he knew the birthdays and those of the deaths, etc., of all his friends and acquaintances."* **Scholarliness:** "Very studious but not large acquirements." [H] **Steadiness:** "Steadiness." [I] *"Continuous pursuit of certain studies from an early age."* **Religious Bias:** *"I gave up common religious belief, almost independently from my own reflection."*

2. **Origin of Taste for Science:** "Certainly innate." [F] *"Certainly innate.... Strongly confirmed and directed by the voyage in the...."*

3. **Education:** "Small schools and Shrewsbury, Edinburgh & Cambridge. No honors. I consider that all I have learned of any value has been self-taught. Educ. restrictive of observation being almost entirely classical. Conducive to health. Merits—None whatever. Demerits—No mathematics, nor modern languages, nor any habits of observation or reasoning." [B] *"No mathematics nor modern languages, nor any habits of observation or reasoning."*

HENRY DEBUS

1. **Qualities: Health:** "Very delicate as a child, afterwards always good. F. Healthy in early life, later rheumatism. f. Do not know; she died when I was young." [D] **Mental Pecularities:** "Steadiness, disinterestedness. M. I am informed that my mother was very clever and had great powers of mind, but I know no details. Relations have often said to me, 'your mother had great intellectual powers.'" [J]

WARREN DE LA RUE

1. **Qualities: Health:** "Excellent up to age of 59. Scarcely any illness. F. Excellent except gout after 40.

f. Excellent." [D] **Energy:** "Physical—'Capable of any amount of bodily exertion if interested. Could work night after night at telescope'; Mental—'Indomitable in pursuit of any study or business,' see his account." [C] **Mental Peculiarities:** "Very prolific in mechanical inventions; a keen experimentalist in chemistry and physics and having steady business habits but never fond of business. A power of concentration of thought to the one object in hand, to the total forgetfulness of others for a time; a determination never to leave unaccomplished a matter once taken in hand; a great love for science, a complete disinterestedness in its pursuit; an appreciation of the work of others, and public spirited. Never a partisan; religious feelings not great; high moral feelings and complete tolerance of the views of other persons. F. A prolific inventor; not by any means a good man of business, but most energetic in it. [18] Disintguished for his steady pursuit of any matter once taken in hand; great curiosity about facts and love of the new; religious feeling not great. M. In household matters a most careful and excellent manager; her social affections were a charming characteristic; religious but not obstructively so. Y.S.B. Myself and two others no ear for music; three others were musical." [J] **Business:** "Steady business habits, but never fond of business." [A] **Independence:** "Self-reliant. F. Self-reliant." [E] **Memory:** "Peculiar memory; bad for things and good for what is connected together." [G] *"A peculiar memory; bad for names of persons, plants, places, etc.; good for subjects connected with others; not bad for numbers. Father—A most marvelously retentive memory; he could relate minute details of historical occurrences, names of actors in politics, almost all he had ever read (he was a great reader), and was in consequence a most lively companion. Mother—Not very good."* **Scholarliness:** "Always a student, but not in literature." [H] **Steadiness:** "Determination never to leave unaccomplished a matter in hand." [I] *"Determination never to leave unaccomplished a matter once taken in hand."* **Religious Bias:** *"Religious feelings not great."*

2. **Origin of Taste for Science:** *"Naturally fond of mechanics and of physical science, in which all my study has taken the direction of those departments bearing on ...,* [19] *owing to my feeling that, through the possession of special instruments for investigations in it, I could work to greater advantage; not from any natural preference for ... over the other departments of physical science."* (identification uncertain)

PHILIP GREY-EGERTON

1. **Qualities: Health:** "Excellent. F. Excellent. f. Excellent." [D] **Energy:** "Physical—'A strong walker;

---

18 Warren De La Rue's father, Thomas De La Rue, was the founder of a large London stationers firm.
19 Astronomy? De La Rue's work was especially connected with the application of photography to astronomy.

never know what it is to be tired.' " [C] *"I have been and still am a strong walker, both mountaineering and deer-stalking. I never knew what it was to be tired, but, after the hardest day, was ready to start again with six hours' sleep. Although in my sixty-seventh year, I am still an indefatigable deer-stalker."* **Memory:** "Retentive for nomenclature, not for numbers or history." [G] *"Retentive for nomenclature, but not for numbers or history."*

2. **Origin of Taste for Science:** "Yes. I was always fond of natural history." [F] *"I was always fond of natural history; collected plants, insects, and birds, at [school] and fossils at [college],*[20] *where . . . 's lectures attracted me to geology, and subsequently, by the acquaintance of Prof. . . . ,*[21] *to the particular branch [of it which I have pursued]."* [22]

3. **Education:** "Small schools & Eton. Swiss private tutor 11–13. Conducive to observation & to health. Merits—a good grounding in Latin, Greek and French. Demerits—omission of mathematics, German & drawing." [B] *Omission of mathematics, German, and drawing."*

FREDERICK J. EVANS

1. **Qualities: Energy:** "Mental—'Far above the average though I hesitate to call it remarkable. For 25 years I never ceased using my brain.' " [C] **Mental Pecularities:** "I consider that I have special talent for the practical business of life, more especially in the application of my acquired knowledge, scientific and otherwise, promptly and effectively in the various duties of my official position.[23] Steady in action and fairly so in thought; tendency towards strong feelings and partisanship rather than the reverse. When once I have formed an opinion, and I take especial pains to form sound ones, this opinion is retained with a dogmatic spirit unpleasant to my feelings; hence my shrinking from inquiry into religious beliefs and receiving pleasure from the investigation of physical subjects. Social affections solid. No love whatever for the marvellous and little belief in new things, by rumour, till they are tested honestly, and by sound principles; curiosity rather feeble that otherwise; love of pursuit warm; constructiveness of imagination feeble; foresight strong; public spirit fairly strong; disinterestedness, have always cultivated and with success." [J] **Business:** "Yes for the practical business of life, administrative power." [A] **Independence:** "Yes in social and pol.; no in religious. I have shrunk from the latter in order to obtain peace of mind and leisure for my studies." [B]

Memory: "Good, far above the average." [G] *"I should say far above the average. I can now refer to note-books of thirty years past and select a special observation. In other words, it is a capital working memory. I never tried to learn pages of poetry, etc.; in this I should probably have failed."* **Scholarliness:** "Yes." [H] **Steadiness:** "A certain tenacity of opinions retained with a dogmatic spirit, repugnant to my feelings." [I] **Religious Bias:** *"I have not cultivated independence of judgment in religious matters; I have shrunk from so doing in order to retain peace of mind and leisure for my varied studies."*

2. **Origin of Taste for Science:** "Yes. I was always attracted to men of ability." [F] *"From youth I always preferred the man of marked ability to the man of action alone. Thrown for so many years of my professional life among men chiefly of the latter class, and my sympathies being more drawn toward those in the decided minority, my tastes were, I conceive, not acquired, but innate. In the early days of my professional career I gained the friendship of . . . , of the highest professional standing, whose acquired general knowledge and love of science and observation were far beyond those of the ordinary . . .*[24] *of his time. I was both his young friend and favorite assistant for three years. He imbued me with his respect for science, and formed my character for earnestness and accuracy. . . . To some extent, my tastes were determined by events after manhood; because in . . . extending over ten years, I held positions of great responsibility [in different parts of the world], but I consider my scientific tastes were formed in youth, that is, from sixteen to twenty-one years of age."*

3. **Education:** "Small schools to aet. 13, then to sea, where life was highly conducive to observation and generally so to health notwithstanding bad and insufficient food. Merits—Steadily taught by one eccentric schoolmaster, accurate spelling, clear, neat and intelligible writing and accurate simple computation. On going to sea aet. 13 I really think I started with the best education I could have had." [B] *"The steadiness with which I was taught, by one eccentric school-master, reading and accurate spelling, clear, neat, and intelligible writing, and quick and accurate computation by all the primary rules of arithmetic. Faults in these several branches were never overlooked, and all competition was for excellence in each; Latin and French were evidently thrown in to please parents. Going to sea at the age of thirteen, I really think I started with the best education I could have had. Compared with my youthful messmates, some of whom had passed through public schools, I was far their superior in writing (I soon acquired chart-drawing and sketching from Nature), and in calculation of the day's work, and in astronomical observations."*

---

[20] While a student at Oxford, Egerton had made collections of fossil fishes.

[21] Egerton studied at Oxford under Buckland and Conybeare.

[22] Paleontology?

[23] Captain Evans was appointed superintendent of the compass department of the navy in 1855; he dealt in this capacity with the problem of using the compass in iron ships.

[24] Seaman?

JOHN EVANS

1. **Qualities: Health:** "Good. F. Good to within a year of death. f. Good." [D] **Energy:** "Physical—'Fair, have walked 40 miles a day'; Mental—'I must leave my works to answer this question.' " [C] **Mental Peculiarities:** "Considerable mechanical and constructive talent, great business aptitude, being quick in decision and observation; steady, the reverse of flighty; not a partisan; remarkably strong curiosity about facts; love of pursuit. F. Very musical; an excellent artist; a good turner. f.F. Aptitude for languages; great collectors of books and coins and objects of natural history.[25] f.M. Mechanical ingenuity." [J] **Independence:** "Fond of writing pamphlets and tract matters." [F] **Memory:** "Very strong both for objects in large collections and for verbal quotations." [G] *"Considerable, both verbal and objective; great facility in quotations; familiarity with large collections of coins and specimens.* Father *and* Mother—*both good memories."* **Mechanics:** *"Considerable mechanical skill."*

2. **Origin of Taste for Science:** *"Decidedly innate as regards coins and fossils. My father and an aunt collected coins and geological specimens, and I have both coins and geological specimens, which have been in my possession since I was nine years old.[26] Subsequently my pursuits were influenced to some extent by the discoveries in . . . ,[27] but at that time I had already a considerable collection."*

3. **Education:** "Schools till aet. 16, was with a tutor in Germany for 6 months. Subsequent technical training and reading. Conducive to observation & to health. Merits—variety of subjects and attention to details. A combination of home and school education, my father having been headmaster of the school." [B] *"Was at school till sixteen years of age, and with a tutor in Germany for six months; after then, technical training and teaching.[28] The education was conducive both to observation and health. Variety of subjects and attention to details. A combination of home and school education, my father having been head-master of the school."*

WILLIAM FAIRBAIRN

1. **Qualities: Energy:** "Physical—'Active & energetic from infancy to 84'; Mental—'I must leave my works to answer this question.' " [C] *"Active and energetic from infancy to eighty-four years of age. In mind —I must leave my works to answer this question; but I believe I have been a hard worker during the whole*

period of my existence. [N.B.—No doubt of it.] Father—*Energetic, both in body and mind; muscular; a great reader.* Mother—*Delicate, but active and intelligent."* **Mental Peculiarities:** "The most prominent are perseverance and industry. I think we are more dependent on these than in relying on the efforts of genius. . . . A willing mind and determination to persevere is in my opinion the most direct road to success; we must however exercise a sound judgment in the selection of subjects on which to exercise our thoughts." [J] **Independence:** "Perfect; F. & f. Perfect." [E] **Memory:** "Good." [G] *"Very good memory as far as my eighty-fifth birthday."* (identification uncertain) **Steadiness:** "Perseverance and industry the most prominent qualities." [I] *"The most prominent are perseverance and industry. A willing mind and determination to persevere is, in my opinion, the most direct road to success; we must, however, exercise a sound judgment in the selection of subjects on which to exercise our thoughts."*

2. **Origin of Taste for Science:** "I cannot tell." [F]

3. **Education:** "Small schools aet. 6–14, and then became my own schoolmaster. I had a thirst for knowledge. Not conducive to observation and acted both ways, as regards health. Merits—good so far as it went." [B]

ARTHUR FARRE

1. **Qualities: Health:** "As a child delicate; as a youth, healthy; as a man never ill, only absent from professional duties 2 days in 30 years; only 2 headaches in my life. F. Much illness early but died old. f. Always healthy." [D] **Mental Peculiarites:** "[Abstract] Very fond of mechanical contrivances; often as a child, when confined by sickness, invented my own toys and childish musical instruments out of combs, jews harps—etc. This strong natural bias for mechanical work has not worn out, but its development has simply been impeded by more serious duties of another kind, and I occasionally indulge in it. Fond of music; regular business habits; my professional life is strictly methodical. In ten consecutive years in earlier life I never took a day's holiday. Every working day is still mapped out into hours, half-hours, and quarters. F. Had no mechanical turn whatever; was certainly very unbusiness like in the sense usually understood by business men (of turning his time to the best account in a pecuniary, not a conscientious, sense). His religious views were decidedly evangelical; fond of listening to music but had no acquaintance with it either theoretically or practically. M. As I was only five years old when my mother died I cannot fill up these questions with any certainty." [J] **Business:** *"My professional life is strictly methodical; every working-day is still mapped out into hours, half-hours, and quarters."* **Mechanics:** *"Very fond of mechanical contrivances. Invented and made my own toys as a child. Mechnical tastes are still largely indulged in intervals of leisure."*

[25] John Evans' father was Rev. Dr. Arthur Benoni Evans, headmaster of the grammar school at Market Bosworth, and a collector of coins.

[26] "When a boy of nine he had shown leanings towards natural science, and had hammered out for himself a collection of fossils from the Wenlock limestone quarries at Dudley." *Dictionary of National Biography.*

[27] Prehistory?

[28] At the age of seventeen Evans entered his uncle's paper-manufacturing business.

WILLIAM FERGUSSON

1. **Qualities: Energy:** "Physical—'Considerable in fulfilling any task, have dissected 16 hours a day in critical times'; Mental—'Considerable. Wrote and superintended 1st edition of Practical Surgery, 200 to 300 woodcuts all without any assistance whatever (during much professional practice) in 7 months.'"[c] *Considerable enduring power in fulfilling any given task or duty; have dissected continually for three or four weeks eight or nine hours a day, devoting some sixteen hours to work at critical times.* In mind—*Considerable. Wrote and superintended first edition of . . . ,*[29] *giving instructions to artists regarding from 200 to 300 woodcuts, correcting press, etc., without assistance, in about seven months [all this in addition to professional work]; hard work for mind as well as body."* **Mental Peculiarities:** "Special love of mechanics; good self-taught cabinet maker, good blacksmith, have made brass bound writing desks, lithotrites, cut models of houses with penknife,, etc.; fair business habits when interested in the work; fond of music; made violins; no taste for mathematics or arithmetic. Determined at 18 years of age to try to excell in surgery; hence the characteristics above referred to, steadiness of severe work and continuous effort; strong friendly feelings; strong social affections, always suspicious of the new and marvellous; special anxiety about facts."[I] **Independence:** "Fairly independent but not obstinate, as a rule."[F] **Memory:** "Great for faces and places."[G] "*Great memory for faces and objects once seen.*" **Scholarliness:** "Always fancied myself idle and dull until I compared my work with that of contemporaries."[H] **Steadiness:** "Determined aet. 18 to try and excell in surgery. Hence to have presumed steadiness."[I] **Mechanics:** "*Special love of mechanics; a good amateur cabinet-maker and blacksmith. Made lithotrites.*"

2. **Origin of Taste for Science:** "I cannot perceive."[F] "*I cannot perceive that they were innate. Possibly my tastes were due to retentiveness of memory as to objects and facts, and a strong impression that good surgery is a great fact. Subsequently, by the approval of teachers, when between the ages of eighteen and twenty, having been selected chief assistant to the most popular teacher of anatomy of his day,*[30] *and also to a professor of surgery.*"

3. **Education:** "Schools & Edinburgh University. Considerably self-taught. Neither conducive nor otherwise in any special way to observation. Yes to health, being as a boy fond of fishing and out of doors. Merits—none. Demerits—Omission of almost everything useful and good except being taught to read—

Latin, Latin, Latin!"[B] "*Omission of almost everything useful and good, except being taught to read. Latin! Latin! Latin!*

WILLIAM H. FLOWER

1. **Qualities: Health:** "Variable, generally good. F. Generally good. f. good."[D]

DAVID FORBES

1. **Qualities: Health:** "As a rule, excellent. F. As a rule, excellent. f. Delicate, consumptive; died of it."[D] **Energy:** "Physical—'Extremely energetic. All life accustomed to extremely rough travel, almost continuous from 1846 to now. Often not 12 or 16 nights in bed. Restless'; Mental—'Considerably so.'"[c] "*Travelling almost continuously from 1846 up to the present time. Restless. All life accustomed to extremely rough travel; often months without house or tent. Of mind—restless. Father—Very energetic; restless. In old age travelled considerably. Mentally restless. Mother—Quiet and delicate.* **Mental Peculiarities:** "Specially interested in applied science of all kinds; determination, persevering (no patience); strong feelings. Believe myself thoroughly disinterested; much religious bias of thought; but no respect for revealed religion as a base for such a bias. F. Thorough practical business man; steady; desire to travel. M. Religious and extremely charitable."[J] **Independence:** "Quite independent." **Memory:** "Excellent; but not for dates or names."[G] "*Bad memory for names and dates, but good as regards facts or circumstances; principles in physical sciences are clearly retained. Father—Excellent memory for historical events, including dates and names in ancient and modern history. Mother—Moderately good.*" (identification uncertain) **Scholarliness:** "Studied over a wide field, too much so, not concentrated enough."[H] **Steadiness:** Determination, persevering."[I] **Religious Bias:** "*Much religious bias of thought, but no respect for revealed religion as a base for such a bias.*"

2. **Origin of Taste for Science:** "In great measure innate."[F] "*I believe the desire for information and habits of observation to be in a great measure innate. They were first developed by a little elementary teaching in physics and chemistry, at school, between the ages of seven and thirteen. I worked alone at science at home, from the age of eleven years, where I was encouraged by the example of an elder brother.*[31] *Subsequently, my pursuits were much influenced by being thrown, at the age of nineteen, on my own judgment and resources. I founded a mining colony in the backwoods of . . . ,*[32] *and had to carry it out with several thousand people, quite alone.*"

---

[29] Fegusson's *Practical Surgery.*

[30] "He became an assiduous pupil of Dr. Robert Knox, the anatomist . . . , who was much pleased with a piece of mechanism which Fergusson constructed, and appointed him at the age of twenty demonstrator to his class of four hundred pupils." *Dictionary of National Biography.*

[31] David Forbes was the brother of the biologist Edward Forbes.

[32] Norway. He was superintendent of a mining and metallurgical works there.

3. **Education:** "Schools—Edinburgh University. Worked alone at science at home from 11 years old. A little elementary Physics and Chemistry taught me aet. 7–13 at one school was most conducive to habits of observation. The teaching of only Latin & Greek at a subsequent school was just the reverse. Health—always strong—never affected by school work. Merits—All I attribute are to those classes of Physics & Chemistry already mentioned. Demerits—Enormous time devoted to Latin & Greek; with which languages I am not conversant."[B] *"All the merits [of my schooling] I attribute to a little elementary physics and chemistry, taught me between the ages of seven and thirteen." "Enormous time devoted to Latin and Greek, with which languages I am not conversant."*

GEORGE C. FOSTER

1. **Qualities: Health:** "Generally good. F. Generally good. f. generally good."[D] **Mental Peculiarities:** "As a boy took great delight in watching machinery and workmen and in making mechanical toys, as water-wheels, wind–mills, carriages driven by clock springs, etc. Discovered accidentally the action of the syphon (at about 12 years old) and thought out nearly the right explanation. Took great pleasure in my first reading of Euclid at school (at 12 or 13) in which I was practically unaided, as the master used to be perplexed by my drawing the figures in ways he was not accustomed to; fond of music, but have never acquired practical skill. F. Has a very good musical ear and considerable skill in vocal music. Son of a working plasterer, he became one of the proprietors of the print works where he was apprenticed as a boy. Devoutly religious and active supporter of the efforts of his own religious body . . . but *extremely* liberal in his judgment of those who are less "orthodox" than himself; has much less sympathy with those who are inclined to stifle rational inquiry. Has been an active promoter of unsectarian education and philanthropic movements. One of Cobden's earliest friends and supporters. For many years one of the leaders of the Liberal party in his own borough (Clithero). M. Like my father, combines strong religious feelings and strict practice with remarkable liberality. Y.B.S. All above average stature . . . all rather shy and retiring. f.F. Uniformly tall and slender; mostly long lived."[1]

2. **Origin of Taste for Science:** *"From an early age I was addicted to mechanical pursuits. In the last few years of my school-days I took to chemistry. Entered . . . college,*[33] *expecting, after two or three years there, to [join a relative's] business as calico-printer, and gave especial attention to chemistry on that account. . . . I had never attended specially to physics until appointed professor of natural philisophy*[34]. *. . .*

---

[33] University College, London.
[34] Professor of natural philosophy at Anderson's University, Glasgow.

[This and subsequent similar advancement[35]*] determined me to devote myself thenceforward definitely to physics, and not to try for a chemical appointment. . . .*

WILSON FOX

1. **Qualities: Health:** "Good. F. Good. f. Fair."[D] **Energy:** "Fair, not remarkable."[C] **Mental Peculiarities:** "Naturally impulsive but steadiness cultivated; love of pursuit; social affections strong. F. Considerable mechanical powers. Impulsive; strong affections; strong religious feelings; love of new; marked foresight; very disinterested; public spirit strong."[1] **Memory:** "Very good as a boy and young man."[G] *"Very good as a boy and young man."* **Steadiness:** "Naturally impulsive."[1]

2. **Origin of Taste for Science:** "No."[F]

3. **Education:** "Schools 5–16. University College, London. Observation—no. Health—indifferent. Merits—Science taught at school (aet. 11–16). Demerits—Want of thoroughness in early teaching."[B] *"Science taught me at school, between the ages of eleven and sixteen." "Want of thoroughness in early teaching."*

DOUGLAS GALTON

1. **Qualities: Health:** "Good. F. Good but gouty. f. Good."[D] **Mental Peculiarities:** "Fair business habit [this is greatly underrated. Ed.]. Some (considerable) mechanical aptitude; no talent for music; fond of the marvellous. F. Good business habits; hated music; steady, not impulsive. M. Liked music; generally un-business-like but answers letters with great rapidity, and keeps account books which she never balances; Very impulsive. M.F. Strong mechanical aptitudes, but irregularly ditsributed."[1]

JOHN H. GLADSTONE

1. **Qualities: Health:** "Good throughout. F. Good till aet. 27; dyspepsia ever since. f. Generally bad."[D] **Mental Peculiarities:** "It is difficult to estimate one's own peculiarities, but I may credit myself with more than the usual amount of those underlined—[religious bias of thought, love of the new and marvellous, curiosity about facts; love of pursuit; constructiveness of imagination,] and with tenacity of purpose, while partisanship is so deficient that I always feel indisposed to follow the lead of others. F. Great tenacity of purpose."[1] **Religious Bias:** *"It is difficult to estimate one's own peculiarities, but I believe I may credit myself with more than the usual amount of (. . . and) religious bias of thought. I have mixed and worked with the Christians of most of the Protestant Churches."*[36]

---

[35] Professor of physics, University College, London.
[36] Gladstone was "keenly interested in Christian endeavour, organising devotional meetings and Bible classes among educated men and women. He was a vice-president of the Christian Evidence Society, and wrote and lectured frequently for it on Christian apologetics." *Dictionary of National Biography.*

2. **Origin of Taste for Science:** "Yes. They date from a very early period and there was little to produce them in my early surroundings." [F] *"They date from a very early period, and there was little to produce them in my early surroundings. As a small boy I was fond of reading books bearing on natural science. I was taught at home with my brothers, and was partially self-taught also. We had always the example of industry, and were encouraged to think for ourselves. I first studied chemistry at . . . College."* [37]

3. **Education:** "No schools. University College, London, 1st class medal, & Giessen. Brought up at home and partially self-taught. Observation—rather so. Merits—We had always the example of industry & were encouraged to observ & think for ourselves. Demerits—perhaps too much shut-out from the company of other boys." [B] *"Being brought up at home; was perhaps too much shut out from the company of other boys."*

JOHN EDWARD GRAY

1. **Qualities: Health:** "In childhood very bad. Since then, good. Deaf & dumb till aet. 10; Very bad; frequent spitting of blood. f. Good, lived to old age." [D] **Energy:** "Physical—'Considerable energy and endurance of fatigue'; Mental—'Considerable & enduring power of brain work.'" [O] **Mental Peculiarities:** "Talent for mechanics and practical business, determination to succeed as far as possible in all my undertakings; particularly devoted to the education and amelioration of mankind in every respect; [38] unceasing exertion in behalf of the advancement of scientific knowledge, especially botany and zoology; have always acted fearlessly. F. Good business habits; self-teaching, for having been born after his father's death and his mother's having been left with only a small annuity was therefore only educated by his mother and himself. Y.S.B. Two brothers besides myself were ardent lovers of natural history. f.F. Uniformly fond of natural history and were nursery and seeds men to four generations at least in Pall Mall. My grandfather introduced grass [? FG] cutting and flour of mustard from Holland and carried on their manufacture in mills at Deptford." [J] **Business:** "Practical business." [A] **Independence:** "Always have acted independently in social, pol., & religious matters." [E] **Memory:** "Very good." [G] **Scholarliness:** "Studious always." [H] **Steadiness:** "Determination always to succeed." [I] **Mechanics:** *"Talent for mechanics."*

2. **Origin of Taste for Science:** "Yes, inherited from my father's family." [F] *"[Yes.] 'Inherited from*

my father's family, who have generally been attached to natural history [especially botany; most remarkable examples are given] my scientific tastes were largely determined by being appointed. . . .'"*

3. **Education:** "No school. No college. Entirely self-taught in all important respects. Taught to read by my mother and occupied by my father in making extracts from books. Conducive to observation & to health, being induced to collect plants & animals—fresh air. Merits—self-teaching; independence." [B]

ROBERT H. GREG

1. **Qualities: Health:** Generally pretty good; better after aet. 50; never very strong, never severe illness, only twice in life 3 days together in bed. F. rather delicate when young, afterwards strong & always well. f. always delicate." [D] **Mental Peculiarities:** "Very methodical; fond of and quick in understanding machinery; [39] excellent business habits; great love of order and instinctive habits thereof. *Utterly* devoid of mathematics and of music. Logical; very particular as to style; but alas! I know nothing thoroughly and never learned anything *au fond.*" [J] **Mechanics:** *"Fond of and quick in understanding machinery."*

WILLIAM GROVE

1. **Qualities: Health:** "Very indifferent; strong frame but weak & torpid liver. F. Delicate when young, robust in appearance afterwards, but always nervous & depressed. f. (No returns)." [D] **Mental Peculiarities:** "Took very early to physical sciences; made electrical machines and air pumps at 12 years old; business habits bad except what is the result of long discipline, but always punctual in appointments. Great want of curiosity as to the ordinary subjects of human gossip, but great love of solving scientific mysteries. Little interest in science as soon as it becomes exact and reducible to number and weight but great interest in solving enigmas such as chemical and electrical phenomena, etc." [J]

2. **Origin of Taste for Science:** *"My first start was reading a child's story called the 'Ghost,' where a philosophical elder brother cures his younger brother of superstition, by showing him experiments with phosphorous, electricity, etc. This set me on making an electrical machine with an apothecary's phial, etc. I was then about twelve years old. My grandfather had scientific tastes to some degree. My grandmother's brother . . . was a good amateur chemist and astronomer. He was a well-known leader of musical, and, to some extent, of scientific society at . . . ."* (identification uncertain)

ALBERT GÜNTHER

1. **Qualities Health:** "Always perfect, except one attack of rheumatism aet. 23 and overwork aet. 36.

[37] Gladstone attended the chemistry lectures of Professor Thomas Graham at University College, London.

[38] "The establishment of public playgrounds, coffee-taverns, and provinicial museums engaged his attention; he was a promoter of the Blackheath Mechanic's Institution, one of the earliest institutions of the kind; he was a strong advocate for the more frequent opening of museums free of charge. . . ." *Dictionary of National Biography.*

[39] Greg was associated with his father's cotton mill in Manchester.

F. Perfect up to time of his early death. f. Perfect." [D]
Energy: "Physical—'When fishing and shooting in holidays, I am the whole day on my legs, never resting'; Mental—'From 1859 to 1871 examined and named 40,000 examples, described 80,000 species, wrote 6,000 pages of print and corresponded also.' " [O] *"When fishing or shooting (my only occupation during the holidays) I am the whole day on my legs. Of mind—in thirteen years I examined and named some 40,000 examples, described about 7,000 species, wrote some 6,000 pages of printed matter, carrying on at the same time a great deal of correspondence. Father—I cannot say. Mother —Is active the whole day. At the age of sixty-three she took sole charge of my child, then but a few weeks old, nursing it for three years, night and day. Energy of mind equal to that of her body."* **Mental Peculiarities:** "Strong business habits and steadiness; very strong feeling against meddling in scientific matters by amateurs or mercenaries; affection for the persons and doctrines of my teachers and old associates in zoology; love of the new and marvellous, love of pursuit; constructiveness of imagination and great disinterestedness. I have a taste for and was formerly a performer of music. A hasty temper is more or less present in all my maternal relations and ancestors as far as I know them. . . ." [I] **Business:** "Business habits strong." [A] **Independence:** "I acknowledge no law but morality; f. shares my views in matters pertaining to her sphere." [E] "In social and political matters I acknowledge no other law but that of morality. For instance, I think it absurd to prevent connexions by marriage from marrying, in a country in which marriages of blood-relations are sanctioned. In English scientific society there is too much respect paid to rank and personal influence at the expense of real acquirement and merit; & whenever I can I counteract this tendency. I feel and act thoroughly independent in religious matters as far as they are instituted by man. I rarely go to church because forms introduced into the English service, have no devotional effect upon me." [M] **Memory:** "Excellent but peculiar; recollection of forms but forget names." [G] *"I recognize most of the animal forms which I have previously examined, but I forget easily the details of their structure, also their systematic names (specific, not generic). Likewise I have a good memory for faces, but not for names of persons; could never remember historical dates."* **Scholarliness:** "Yes." [H] **Steadiness:** "(Puts himself F to nearly all the qualities.)" [I] **Effect of religion on science:** "Not a deterring effect, but it acted as a guide." [M] *"Not a deterrent effect; but it acted as a guide.."*
2. **Origin of Taste for Science:** "My love for animals was very strong in my early youth. I should have been an observer of animal life under any conditions." [F] *"How far innate, and how far acquired and developed from my early youth, I cannot say. My love for animals of all kinds was very strong, and to gratify it I overcame*

every obstacle put in my way at home, when I was a boy. I trace the origin of my interest in science to the earliest impressions of my childhood, all of which, so far as I recollect them, are connected with my father and the various animals he brought me as pets. They were not largely determined by events after manhood. I should have been an observer of animal life under any conditions under which I might have lived."
3. **Education:** "½ year elementary school; 10 years gymnasium; Tübingen, 5 years, Berlin, 1 year, Bonn, 1 year; Tübingen, 1½ years. Two medals at school for general proficiency. Gained a scholarship for 6 years before entering the university; passed the examination; obtained M.A. Theology & Ph.D. degrees cum laude; passed the 1st and 2nd examinations for medicine and surgery, was made an M.D. for scientific labours. No teaching at home, but every day preparing for next day's school. To observation—Very restrictive. To health—much the reverse. Merits—A steady and systematic application to work first forced, later habitual. A good basis of the elementary knowledge of a variety of subjects with a thorough training in some. A conscientious supervision of my studies (and of my life) during my first four years University life by the authorities of the College to which I was attached. During those 4 years my zool. studies were entirely a work of superogation, & I became all the more devoted to a science which I then cultivated under difficulty. Demerits—At an early age being destined to the Church, I consented to it before my own judgment was formed & I consider the 3 years devoted to theology study almost as loss of so much time of my life." [M]

WILLIAM GUY

1. **Qualities: Health:** "Slightly marked with scrofula. F. Neuralgic; tic doloreux for year. f. Consumptive." [D] **Energy:** "Physical—'Active; can endure fatigue; Mental—'Can get through a good deal of work'." [O] **Mental Peculiarities:** "None for music; love of figures and tabulations [40] (the same love was remarkable in my paternal grandfather. I inherit many mental peculiarities and talents from him, none from my father). [J] **Independence:** "In all." [E] **Memory:** "Bad, especially for names." [G] *"A bad memory, especially for names."* **Scholarliness:** "Certainly." [H]
2. **Origin of Taste for Science:** "I think innate." [F] *"Innate, I think. I inherit many mental peculiarities and talents from my paternal grandfather, among which is a love of figures and tabulation; none from my father. I cannot [otherwise] trace the origin of my interest in science, nor were my tastes largely determined by events after manhood."*
3. **Education:** "7–15 school. Cambridge — low wrangler. Christ's Hospital. Restrictive to observation. Not conducive to health. Merits—none. De-

---

[40] Guy was particularly concerned with the preparation of various tables as a member of the Statistical Society of London.

merits—Latin through Latin; nonsense verse—want of games, a solitary chold among old people."[B] "*Latin through Latin—nonsense verses.*"

AUGUSTUS HARCOURT

1. **Qualities: Energy:** "Very untiring. Often pass from a walk to a run, even in the streets."[O] "*I seem to possess the same unweariedness as my father, and find myself trotting in the streets as my father used to do. Father—Was very untiring; he tells me he has ridden one hundred miles in a day. He could walk up one of the North Wales hills when nearly seventy, and used to go long distances in London, passing often from a walk into a run.*" **Mental Peculiarities:** "Chemistry was my earliest taste and the favorite occupation of my holidays. F. My father was never much of a reader, but his chief intellectual tastes were scientific. In old age the reading of a scientific book or listening to a scientific lecture interests him more than any thing else. Y.S.B. My brother and one of my sisters have the same scientific pursuits[41] as myself. f.F. Two of my father's brothers made chemistry their pursuit; and at least two more are regular attendants at the R. Institution lectures."[J] **Memory:** "Very treacherous in business; retentive of verse."[G] "*My father and myself have memories of the same character: treacherous in matters of business, and very retentive of scraps of verse read over and learned long ago. When my father was to have met me, a little boy returning from my school at the end of the half, he would forget all about it. My engagements sometimes suffer . . . from similar forgetfulness.*" **Scholarliness:** "Chemistry was the favorite occupation of my holidays."[H]

2. **Origin of Taste for Science:** "Thoroughly."[F] "*Thoroughly innate. My first taste for chemistry dates from the possession of a chemical box, when I was a little boy. Whenever I had a chance of turning from other studies to natural science, I always turned. I liked play better than all other work, and chemistry better than play.*"

3. **Education:** "2 schools—a little science, then Harrow & Oxford where a college laboratory at Balliol then almost untenanted was available—1st class nat. science.[42] Demerits—not taught mathematics nor any natural history science, to which I should have taken con amore."[B] "*Not taught mathematics, nor any natural science, to which I could have taken con amore.*"

WILLIAM P. HATHERLY

1. **Qualities: Health:** "Very good. F. Very good. f. Very good."[D]

2. **Origin of Taste for Science:** "*My interest in science was due to my having been officially employed in the early part of [my career, in a very important statistical inquiry].*"[43] (identification uncertain)

3. **Education:** "Small schools, then at Geneva, natural history under De Candolle. Cambridge 24th wrangler, two unwell for classical exam but afterwards fellow of Trinity College."[B]

SAMUEL HAUGHTON

1. **Qualities: Health:** "Delicate up to 16, since then good, though often had fever and cholera. F. Always good, now aet. 86. f. Good but neuralgic."[D] **Energy:** "Physical—'Considerable when young; a good fencer, swimmer and jumper, walk well'; Mental—'Considerable; have worked hard with the brain for the last 35 years. Mind always at work when awake.'"[O] "*I possess considerable bodily energy, and when young excelled in fencing, swimming, and the high jump. In mind—Have worked hard with my brain for the last thirty-five years, almost without intermission. Father—Considerable bodily energy, and a good pedestrian. Mother—Sluggish bodily powers, but in mind most energetic when once roused to action by a subject that interested her feelings.*" **Mental Peculiarities:** Applied mathematics, oratory. I am a very ready speaker and have a quick sympathy with others. When my feelings are excited I can readily move an audience of 2000 persons and make them laugh or cry (a large audience is much more under the control of a skilled speaker than a small one, possibly, perhaps on account of the larger mass of those who feel and do not think.) Tenacity of purpose and steady industry; intense desire of knowledge of every kind; great sympathy with children and animals; strong religious feelings; lively imagination and keen sense of the ludicrous. My friends consider me disinterested and public spirited. F. Has a lively imagination and strong sense of the ludicrous. M. Extreme tenacity of purpose, which no amount of repeated failures could ever turn aside. Y.B.S. Tenacity of purpose is strongly marked in 5 out of 8 brothers and sisters. These 5 possess also lively imaginations and keen sense of the ludicrous. f.F. Lively imagination and keen sense of the ludicrous. f.M. Tenacity of purpose."[J] **Independence:** "Strong judgments, but care little for politics. Was baptised being born of Quaker parents at my own request aet. 13 and chose my clergyman's life in preference to a good living in commerce; f. of masculine mind—clever manager."[F] **Memory:** "Verbal memory bad."[G] **Steadiness:** "Tenacity of purpose."[I] **Religious Bias:** (history of his fellowship)[K] "*Strong religious felling. My intention on entering . . . was to devote myself to a missionary life in China; but my unexpected success in . . . persuaded my friends to induce me to abandon*

[41] See Galton's discussion of the Harcourt family, *English Men of Science*, pp. 50–51.

[42] "When (Sir) Benjamin Collins Brodie . . . came to Oxford as professor of chemistry in 1855, the Balliol laboratory was placed at his disposal, and Harcourt became first his pupil and then his assistant." *Dictionary of National Biography.*

[43] Lord Hatherly (William Page Wood) was early involved in railway inquiries in testimony before Parliament.

my purpose, on the ground that I might serve God better in my new sphere at home." (identification uncertain).

2. **Origin of Taste for Science:** "As far back as I remember I loved nature and desired to learn her secrets." [F] *"As far back as I can remember, I loved nature and desired to learn her secrets, and have spent my whole life in searching for them. While a schoolboy, I taught myself botany, chemistry, etc. . . . under great difficulties. I had no teacher except a kind apothecary. whose knowledge was limited."*

3. **Education:** "Large schools 11–17. Trinity College, Dublin—many honors & mathematics fellowship. While a schoolboy taught myself Botany, Chemistry, etc. under great difficulties. To observation—no. Health—yes, my open air rambles after plants, my health was otherwise delicate. Merits—well taught in classics and mathematics. Demerits—If possible it should have afforded facilities for the study of the science of observation but I doubt the practicability of this at school." [B] *"Well taught in classics and mathematics. If possible, my education should have afforded facilities for the study of the science of observation, but I doubt the practicability of this at school. While a schoolboy I taught myself botany, chemistry, etc., under great disadvantages.*

JOHN STEVENS HENSLOW

1. **Qualities: Health:** "Excellent. F. Good. f. Good." [D] **Mental Peculiarities:** "Extremely ingenious in devising modes of preserving, exhibiting explaining etc. objects of natural history, in fact he possessed remarkable powers of contrivance and invention. Remarkable steadiness; great power of observation in every branch of natural history; very public spirited and thoroughly disinterested and unselfish. F. Good man of business; extremely fond of birds and other animals and of his garden; at one period had extensive aviaries, many books on natural history, etc. M. Good housewife and needle woman, not much of a reader; extremely warmhearted, affectionate and unselfish; a great admirer and collector of natural and artificial curiosities. F.f. Energy and perseverance decidedly, and in some, cleverness of invention and contrivance. M.f. Energy; decidedly clever and ingenious." [J] **Mechanics:** "[*Was extremely ingenious in devising modes of preserving and exhibiting objects of natural history*]."

2. **Origin of Taste for Science:** "[*A posthumous account.*] 'He appears to have been attached to natural history all his life through, but never took up botany to any extent till the professorship was vacant. [There is some conflict of testimony here.]* [44] *I think his scientific tastes were innate. I have excellent draw-*

ings of insects made by him as a school-boy; also, he made a model of a caterpillar; tried a little chemistry; made lace bobbines of his own contriving. It was said, "Nothing escapes that boy's eyes."* '"

JAMES HEYWOOD

1. **Qualities: Independence:** "[Refers to motions he made in the House of Commons, in advance of popular opinions.]" [E]

2. **Origin of Taste for Science:** "Prof. . . . 's lectures on geology were the origin of my interest in that science; [45] the work of the . . . statistical society [46] in educational inquiries influenced my taste for statistical science; frequent attendance at meetings of the British Association encouraged my scientific tastes."

3. **Education:** "Small schools—1 year at Geneva. Edinburgh University. Cambridge—Senior optime. Conducive to health. Merits—Its thoroughness. Demerits—Greek and Latin were more insisted on than modern languages." [B] *"Latin and Greek were more insisted on than modern languages."*

ROWLAND HILL

1. **Qualities: Health:** "Good. F. Variable. f. Very good." [D] **Energy:** "Physical—'One of the best runners and leapers at school; a persistant walker'; Mental—'Yes, a large amount of brain work.'" [C] *"I was in youth and early manhood bodily active, a good runner and leaper, excelling almost all my schoolfellows [the school was a large one] in both points, and a persistent walker. In mind—During the best fifty years of my life I went through a large amount of brain-work, and vigorously pursued the several interests indicated in the enumeration of my several occupations. Father—In bodily activity much like myself, with the addition that he was a good swimmer. In mind—Capable of great occasional exertion rather than of sustained effort. Mother—In mind—very energetic within a limited range. Always showed great courage, fortitude, and equanimity. In her nursing duties, whether of young or old, was active, persevering, and remarkably successful."* **Mental Peculiarities:** "I had even from childhood a fondness and aptitude for mechanical construction and from boyhood the same for organization. (Numerous, most striking instances given)—Remarkable as regards mental calculation, . . . in particular I became quick and accurate in mentally extracting to the nearest integer the cube roots of large numbers, any indeed not exceeding two thousand millions (Numerous important mechanical inventions and instance of organizing power in manhood are given.) [47] I think my power of fore-

---

[44] In 1825 Henslow resigned the chair of mineralogy at Cambridge to become professor of botany, but biographical accounts indicate that he already had a previous interest in botany.

[45] Heywood was a member of the Geological Society and in 1876 edited O. Heer's *Primaeval World of Switzerland.*

[46] Together with his brother Benjamin Heywood, James Heywood was a founder of the Manchester Statistical Society.

[47] "As a young child he had, on account of an affection of the spine, to maintain a recumbent position, and his principal

sight was always strong; by exercise it eventually became so great as to enable me in the extensive and sometimes perfectly novel operations I was called on to undertake, to provide for almost every contingency.[48] I always took great interest in scientific discovery, mechanical and social improvement and in public affairs generally; was strong in my opinions and to these I gave such effect as circumstances allowed. Under the head of distinterestedness, I twice made large sacrifices of income in order to carry out postal reform. F. Impulsive; strong in feelings and partisanship; also in social affections. Greatly interested in scientific discovery; took great interest in public affairs up to the time of his death; courageous in maintaining his views. M. Had considerable aptitude and I apprehend far greater potentiality for practical business; however, she successfully managed the domestic part of a (very) large household, showing herself especially valuable in times of sickness. Strong in social affections; took great interest in public affairs; courageous in expressing her convictions. Y.S.B. In me and one or two others of my generation the sense of smell, taste and touch were unusually strong; an inheritance from my mother, who for instance could not touch brass without annoyance. I myself have repeatedly succeeded when blindfolded in correctly arranging the pieces on a chessboard, distinguishing the reds from the whites; that is the stained from the unstained. I have also repeatedly performed on a mahogany table, that balancing of an egg, which Columbus according to the story only achieved by cracking the shell. My sight was once so keen that I could take up a single mite from a quantity of cheese-dust. While these powers have faded with age, my taste and smell remain unusually perceptive. . . ."[J] **Business:** "Clearly for business."[A] **Independence:** "F. a liberal when liberalism then styled Jacobinism was highly obnoxious; an early denouncer of slavery & advocate of religious liberty; a freethinker when the world was protectionist."[E] *"My father was a Liberal when Liberalism (then styled Jacobinism) was highly obnoxious, an early denouncer of slavery and advocate of religious liberty, a free-trader when the world was protectionist, and an opponent of unrighteous war when war was most popular. He was for mitigating our criminal code when hanging was regarded as the sheet-anchor, and, in a word, was poltically and socially a very independent spirit."* **Memory:** "Moderate verbal, strong in facts and figures."[G] *"Of moderate verbal memory, but strongly retentive of facts and figures so far as they are related to any subject on or in which I was engaged. Father—Memory very retentive, but not systematic. He had a great amount of information, but had not great acquirements; his familiarity with Scrip-*

ture was, however, remarkable. Mother—Very retentive for small facts and figures." **Scholarliness:** "Yes, but have wanted leisure."[H] **Mechanics:** *"I always took great interest in mechanical improvement."*

2. **Origin of Taste for Science:** "I cannot distinguish."[F] *"I cannot distinguish between what I may have derived from nature and what I may have acquired from intercourse with my father and certain of his friends. When I was eleven years old, my father gave a series of lectures on electricity, mechanics, astronomy, and pneumatics, to all of which, but especially to the last, I paid delighted attention. I presently began to construct apparatus for myself. Subsequently practice in teaching led me to seek for knowledge. Intercourse with men of higher attainments became a great spur; my turn for . . .[49] was favored by my opportunities as an early member of the . . . society."* [50]

3. **Education:** "My father being a schoolmaster,[51] I was at some sort of schoolwork nearly all my life. Principally self-taught though with assistance. On account of ill health I began late so that at 7 I could not read. To observation—yes. To health—yes, rather than otherwise. Merits—Being led to speak freely, to engage in domestic discussions of general policy. I had also early access to tools & materials. Demerits—The omissions which commonly attach to straightened means."[B] *"My father being a school-master, I was at some sort of school-work nearly all my life, but from the age of twelve I was occupied more in teaching than in learning. My education included the various subjects usually taught in English schools, with something of astronomy, pneumatics, electricity, and mechanics. I learned much in conversation with my father, which chiefly took an instructive form. Was led to think and speak freely; also to engage frequently in domestic discussions on questions of general policy. I had also early access to tools and materials."*

THOMAS HIRST

1. **Qualities: Health:** "Rather feeble in childhood, robuster as I grew up, but still liable to great reverses on excitement. F. Remarkably robust. f. Never robust, yet got through much labour & anxiety and d. aet. 77."[D] **Energy** "Mental—'From aet. 20 to 23 frequently worked hard [12 hours a day] but health suffered.'"[C] **Mental Peculiarities:** "Practical business habits good, great love of music and mathematics. Steadiness decidedly marked. Bias towards freedom of thought in religious matters; love of mathematical pursuits predominant; imagination constructive in geometrical matters. F. Devoted to music and fond of elementary mathematics. M. Musical talents fair." **Independence:** Yes, in early life and in religious matters

method of relieving the irksomeness of his situation was to repeat figures aloud consecutively until he had reached very high totals." *Encylopaedia Britannica,* 11th ed.

[48] Hill was inventor of the penny postage and adviser to the postal system.

[49] Statistics.

[50] Statistical Society of London.

[51] Hill's father was a noted educational reformer and the founder of the school at Hilltop, Birmingham.

especially; F. I believe so in him also (but was too young to judge)." [E] **Memory:** "Not retentive." [G] **Scholarliness:** "Decidedly so." [H] **Steadiness:** "Steadiness decidedly marked." [I] *"Steadiness decidedly marked."*

2. **Origin of Taste for Science:** "My taste for mathematics appears to be innate." [F] *"My taste for mathematics appears innate. As a boy, I delighted in sums. I trace the origin of my interest in general science to my acquaintance with . . . ,*[52] *which dates from the time when I was about fifteen years of age. I taught myself in mathematics and chemistry during my apprenticeship to a civil engineer and land-surveyor, and subsequently studied . . . [abroad].*[53] *My scientific tastes were largely developed through my first going [to the continent] with . . . ."*[54]

3. **Education:** "Schools—Universities at Marburg—Berlin & Paris. Self taught in mathematics & chemistry during my apprenticeship to a Civil Engineer. Not to observation, but to habits of contemplation. Not to health during university education. Merit of fostering a love for original research." [B]

JOSEPH DALTON HOOKER

1. **Qualities: Health:** "Very good throughout. F. Very good. f. Very good." [D] **Energy:** " 'Much endurance of fatigue and hard living, excellent mountaineer, travelled much in various climates'; Mental—'An enormous correspondence on a variety of technical and scientific subjects [in addition to high professional success, FG].' " [C] *"When young, and to the age of thirty or more, worked habitually till two and three A.M., often all night. Travelled much in various climates. Much endurance of fatigue and hard living—[an excellent mountaineer]. Of mind—[has risen to the highest position in his branch of science, and conducts an enormous correspondence on a variety of technical and scientific subjects]. Father—Very considerable energy both in body and mind. Mother—Below the average in bodily energy, but remarkabley active mentally."* **Mental Pecularities:** "Very good practical business habits; some tastes for drawing and music but the latter was never pursued. Steadiness, foresight and public spirit, well marked. F. Languages; business habits; music all very strong. M. Love of ruling and an anxiety to convert others to her opinion. f.M. Remarkable artistic and musical power, very irregularly developed." [J] **Independence:** "Habitually challenges conventional ideas." [E] **Memory:** Great power of recollecting forms and points of objective interest, none of numbers and abstract arguments. Languages and poetry soon lost." [G] *"Great power of remembering forms and points of objective interest; none of numbers and abstract arguments. Languages, poetry, etc., soon*

lost if not kept up." **Scholarliness:** "Great." [H] **Steadiness:** "Steadiness." [I]

2. **Origin of Taste for Science:** "In born." [F] *."'My scientific tastes were inborn' [and strongly hereditary]."*[55]

3. **Education:** "Glasgow upper school & University. Occasional tutors at home. To observation—neither way. To health—neither way. Merits—Those of a Scotch education generally. Demerits—Want of training in habits of observation, & the 'teaching' of 110 boys by one master." [B] *"Want of training in the habits of observation." "Two cases: both [being Englishmen] praise Scotch system of education."*

GEORGE HUMPHRY

1. **Qualities: Health:** "Delicate; inflammation in lungs aet. 24, 30, and 36. Typhoid fever aet. 26. F. Delicate but good. f. Good." [D]

2. **Origin of Taste for Science:** "Not at all." [F] *"Not at all especially innate. I could have taken to any other subject quite as well, so far as I know. I trace the origin of my interest in science to the knowledge that I must do my best in it to earn a livelihood and to please my parents. I did not follow my own branch from any special liking—indeed, I disliked it; but it was necessary to follow some branch. The connection with a hospital and medical school in . . . has been an inducement to continue work, and all my life I have worked pretty steadily."*

3. **Education:** "Schools—London University. Cambridge (gold medal for anatomy). Early education at home by mother and father, good. To observation—neither way. To health—rather the reverse. Merits—Rather strict training and being carefully looked after by my father and expected to do my best. Demerits—Perhaps was worked rather too hard at first school." [B] *"Careful and good early education at home by my mother and father; then, rather strict training by my father and by my first school-master. Being carefully looked after by my father and expected to do my best."*

THOMAS HENRY HUXLEY

1. **Qualities: Health:** "As a boy good to 13, then nearly died being poisoned at the first post-mortem I attended. Subsequently incessantly suffered from neuralgic dyspepsia etc. F. Good, d. 77. f. Good on the whole but 'creaking,' but no definite ailment." [D] **Mental Peculiarities:** "Strong natural talent for mechanism, music and art in general, but all wasted and uncultivated. Believe I am reckoned a good chairman of a meeting. I always find that I acquire influence, generally more than I want, in bodies of men and that administrative and other work gravitates towards my hands. Impulsive and apt to rush into all sorts of

---

[52] Probably John Tyndall.
[53] Marburg, Paris, and Berlin.
[54] John Tyndall.

[55] Hooker's father was Sir William Jackson Hooker, Regius Professor of botany at Glasgow.

undertakings without counting cost or responsibility; love my friends and hate my enemies cordially; entire confidence in those whom I trust at all and much indifference towards the rest of the world. A profound religious tendency capable of fanaticism, but tempered by no less profound theological scepticism. No love of the marvellous as such; intense desire to know facts; no very intense love of my pursuits at present, but very strong affection for philosophical and social problems; strong constructive imagination; small foresight; no particular public spirit; disinterestedness arising from an entire want of care for the rewards and honour most men seek; vanity too big to be satisfied by them. F. A good musician and possessed a curious talent for drawing heads with pen and ink. Impulsive but kindly; nothing otherwise remarkable. M. Very impulsive, a strong partisan, strong affections, marked religiosity and a constructive imagination worthy of a novelist. Physically and mentally I am far more like my mother than my father. Y.S.B. All but one had good ears for music and are above the average level in ability; one sister a remarkably good musician, one brother a man of considerable administrative power but broke down in middle life. Family generally;—hot temper and tenacity of purpose; considerable power of expression in writing and speaking." J **Business:** *I believe I am reckoned a good chairman at public meetings, and I always find that administrative and other work gravitates toward my hands.* **Mechanics:** *" 'Strong natural inclination toward mechanism.' [His present profession was accidental and against the grain.]"* **Religious Bias:** *"A profound religious tendency, capable of fanaticism, but tempered by no less profound theological skepticism."*

2. **Origin of Taste for Science:** *"As a boy, I had no taste for natural history, but a passion for mechanical contrivances, physics ,and chemistry. I earnestly desired to be an engineer, but the fact that I had a . . . [near relative]a medical man led to my being apprenticed to him, and I took to physiology and anatomy, as the engineering side of my profession. [The inclinations above mentioned were] altogether innate, and, so far as I know, not hereditary, neither of my parents nor any of the family showing any trace of the like tendencies. My appointment to the surveying-ship . . .[56] made me a comparative anatomist, by affording opportunities for the investigation of the structure of the lower animals. My appointment to . . .[57] forced me to paleontology."*

JOHN G. JEFFREYS

1. **Qualities: Health:** "Rather delicate when young, better afterwards, now good. F. Good. f. Good.[D] **Mental Peculiarities:** "Practical and methodical business habits; steady love of pursuit. M. Musical talent,

especially for singing." J **Business:** "Practical and methodical business habits." A **Scholarliness:** "Studious and persevering." H **Steadiness:** "Steady love of pursuit." I

2. **Origin of Taste for Science:** *"I do not consider them innate, but induced by the following circumstances: when I was at school (between thirteen and fifteen years of age), a lady, and old friend of my mother, gave me a few British shells, with their names, and a copy of 'Turton's Conchological Dictionary.' I thenceforth diligently collected British shells, and afterwards extended my researches."* (identification uncertain)

3. **Education:** "Swansea. Schoolfellows, Bruce—Grove—2 Dnidlmems. To observation—no. To health —no. Demerits—Want of logical and mathematical training." B (spelling of "Dnidlmens" uncertain) *"Want of logical and mathematical training."*

FLEEMING JENKIN

1. **Qualities: Health:** "Good but not strong when young. F. Good. f. good." D **Mental Peculiarities:** "Drawing, considerable as a talent, not as a result. Mathematics. M. Music, much curiosity about facts and love of pursuit. Considerable energy in the members of her family." J

2. **Origin of Taste for Science:** "Decidedly." F *"Decidedly innate. The science of . . .[58] was well taught at the University of . . . ,[59] where I studied, between sixteen and eighteen years of age, and accidentally this became serviceable to me when employed as an engineer by . . . .[60] The friendship of . . .[61] materially affected my career. My tastes were not largely developed by events occurring after manhood."*

3. **Education:** "Schools—13–16 abroad, 16–18 University of Genoa, natural philosophy and science well taught—silver medal. Self-taught in all higher mathematics, mechanics & professional work. To observation—neither way. To health—neither way. Merits—Fairly introduced to many subjects of interest. Demerits—Want of system." B *"Early introduced to many subjects of interest." "Want of system."*

WILLIAM STANLEY JEVONS

1. **Qualities: Health:** "Very good till 30, when from excessive work or other causes it seriously declined. F. Very good throughout. f. Died aet. 50 after several years severe illness. I do not know of anything amiss in her early life." D **Energy:** "Amount of brain work never remarkable." O *"No remarkable energy of body. In mind—never capable of a large amount of brain-work; for years have regarded myself as defective*

---

[56] H.M.S. Rattlesnake.
[57] The Geological Survey.

[58] Natural philosophy.
[59] Genoa.
[60] Newall's submarine cable works.
[61] William Thomson?

in brain-power. [*The actual performance of this correspondent is considerable, and of a very high order.*] Father—*In early life fond of athletic sports, and an enthusiastic sportsman. Energy of mind very remarkable, shown in early university and professional life and all subsequent occupations. He wrote a large number of publications on subject of . . . and . . . controversy.*[62] Mother—*Energy of mind remarkable; zeal in pursuit of interests, excessive.*" **Mental Peculiarities**: " 'A special talent for music, quick and correct ear from his earliest childhood. Plays both piano and organ well. Continuous pursuit of certain studies from an early age. Much curiosity. I often feel a positive pain in passing any object of which I do not comprehend the meaning and construction.' F. An excellent man of business and of great common sense; he had also a strong interest in mechanical matters and was intensely interested in new schemes. He had a remarkable calm, well balanced judgement and temper, and peculiarly conciliatory manners; he seldom or never had any enemies and was not unfrequently asked to mediate in disputes. M. Fond of music and painting. Decidedly strong feelings, partisanship and social affections. Great religious bias of thought; strong love of pursuit and warm public spirit as shewn by (specified philanthropic efforts.) Y.S.B. Music—but considerable differences in other respects which I have often thought could be accounted for on hereditary grounds. f.M. Many uniformities, the chief characteristics being cultivation and refinement of mind joined to want of energy and bodily strength." [J] **Mechanics**: "*I often feel a positive pain in passing an object of which I do not comprehend the meaning and construction.*"

2. **Origin of Taste for Science**: "Innate and developed afterwards."[F] "*I should be much inclined to think there was an innate tendency, but that the tastes were developed by a good and for the most part suitable education. When at my first school, aged ten and a half to twelve years, the head—master gave very clear occasional lessions in moral and economical subjects. I can remember vividly to the present day the impression which those lessons made upon me. As I am not aware that the other boys in the class were equally impressed, I think I must have had an innate interest in those subjects; but the lessons probably increased the interest very much.*"

3. **Education**: "Schools. University College, London. Various Medals—chemistry, botany, logic, political economy. Mother encouraged botany & got prize though only self-taught. To health—neither way. Merits—Yes, education well balanced, omitting three years taught at a private school. Demerits—12–15 at a private school. Latin and Greek grammar, a blank waste of time."[B] "*My education was well balanced; it was general and of a very complete kind, including chem-*

istry, botany, logic, and political economy; but three years (between the ages of twelve and fifteen) spent in learning the Latin and Greek grammars were a blank waste of time.*"

T. RUPERT JONES

1. **Qualities: Health**: "Good throughout. F. Good until last illness. f. Good until an accident."[D]

2. **Origin of Taste for Science**: "*A natural taste for observing and generalizing, developed by noticing the fossiliferous rock which happened to occur in the neighborhood of the school where I was.*[63] *Afterwards the surgeon to whom I was articled, who had an observant mind, fostered my tastes.*" [64]

3. **Education**: "Several small schools, one large, 100 boys, where I was alternately 1st-2nd. Apprenticed to a surgeon. No systematic education, much desultory reading. Not conducive to observation. Merits—Studying Latin & Greek seemed to give a command over words and phrases—learned little at school. Demerits—Want of mathematics and modern languages." [B] "*Want of education of faculties of observation; want of mathematics, and of modern languages.*"

WILLIAM LASSELL

1. **Qualities: Health**: "Remarkably good. F. Uninterruptedly good till fever aet. 45, of which he died. f. Good throughout life d. aet. 75."[D] **Energy**: "Physical—'Remarkable energy and considerable muscular power'; Mental—'Vigorous pursuit of interests. Great industry.' "[C] **Business**: "Practical Business."[A] **Independence**: "Strong independent judgment."[E] **Steadiness**: "Steadiness."[I]

2. **Origin of Taste for Science**: "Entirely innate."[F] "*My tastes are entirely innate: they date from childhood.*" [65]

3. **Education**: "Schools 8–15. A good deal self taught."[B]

JOHN LAWES

1. **Qualities: Health**: "Delicate in childhood & youth. Since then, very good health. F. Severe gout; d. of it aet. 54. f. Exceedingly good health up to now, aet. 81."[D] **Energy**: "Mental—'Very great.' "[C] **Mental Peculiarities**: "For games and sports requiring skill and thought; for figures as bearing on any subject; for music in the form of memory for tunes and for practical business. Steady and intense perseverance; power at will of concentrating the mind on one subject or of

[62] Jevons' father, Thomas Jevons, was an iron merchant who wrote on legal and economic subjects.

[63] Jones was partly educated at Ilminster, "where he began to take interest in geology, collecting ammonites and other fossils from the stone-beds of the upper and middle Lias." *Dictionary of National Biography.*

[64] Probably Dr. Joseph Bunny at Newbury, Berkshire.

[65] "At the age of four or five he amused himself by polishing lenses." *Dictionary of National Biography.*

bringing the mind to bear rapidly on a succession of subjects—great religious bias of thought; mechanical; disinterested. F. Mechanism and games of skill such as whist and chess. M. Mechanism as shewn in the love of building. Impulsiveness, strong feelings, partisanship, social affections and religious faith; disinterestedness." **J** Business: "Practical business." **A** Independence: "Yes, remarkable; F. Decided." **E** Memory: "Remarkable for melodies, not for things." **G** Scholarliness: "Great idleness when young." **H** Steadiness: "Steady and intensive perseverance." **I** *"Steady and intense perseverance."* Mechanics: *"Ground, polished and silvered a 7-inch speculum, and mounted it equatorially."*[66]

3. Education: " 'Schools & Eton—did nothing, always in scrapes.' [Quite self-taught.]" **B**

JOHN LUBBOCK

1. Qualities: Health: "Not strong up to 25, good after. F. Not strong in early life, subsequently good health. f. Never very strong." **D** Energy: "Physical—'energetic [remarkably so, FG]'; Mental—[enormously so, FG]." **C** Mental Peculiarities: "[Exceptionally good business habits; but no returns under this head have been sent, Ed.] F. Mathematics. Y.S.B. Bodily activity." **J** Business: "Clearly high business habits." **A**

2. Origin of Taste for Science: "Innate and I believe hereditary, being derived from my father's father." **F** *"Love of observation and natural history innate; [I had them] as early as I can remember. My grandfather was very fond of natural history, and a [more distant] relative has written an excellent fauna of . . . . The help of Mr. . . .*[67] *has aided me immensely, but not, I think, altered my tendency."*

3. Education: "A good deal self taught. Conducive to health. Merits—Parental encouragement. Demerits —Absence of any scientific training; too much confined to classics." **B** *"Absence of any scientific training; too much confined to classics."*

ROBERT MAIN

1. Qualities: Health: "Unexceptionally good. F. Unexceptionally good. f. Generally good but a rather delicate organization." **D** Energy: "Physical—'Considerable power of enduring fatigue'; Mental—'Considerable mental energy. Self-taught in very scanty intervals of leisure. Can read the chief European languages.' " **C** Mental Peculiarities: "Probably the following characteristics I have beyond the average:— Steadiness of determination even when the subject is

distasteful; strong social and domestic affections, and instinctive religious bias, curiosity about facts; disinterestedness probably considerable. Y.S.B. Mathematics." **J** Independence: "Yes, have taken considerable pains in investigating social, political and religious matters." **E** Steadiness: *"I have, probably beyond the average, steadiness of determination, even when the subject is distasteful."* Religious Bias: "Strong religious bias—(theological books)" **K** *"Instinctive (or original) religious bias, though this may be in part due to early training. . . . I take considerable pains in the investigation of religious matters, one of my amusements being the collection of a considerable theological library, with the books of which I am familiar."*

2. Origin of Taste for Science: *"I am not aware of any innate taste for science. I can only remember in boyhood the influence of the Philosophical Society of . . . , and of a juvenile philosophical society in which I took interest. My interest in astronomy, especially, was very small indeed until I was appointed [to the directorship of an observatory.]*[68]*"* (identification uncertain)

3. Education: "Schools (under an excellent teacher, Mr. Neasy at Northwalk.) Cambridge, 6th wrangler. Also much self-taught. Observation—admirably taught 13–16½ to reason, use my own mind and depend on myself. Health—tolerably healthy—gymnastics. Merits—The habits taught me of using my own mind & of acquiring for myself large masses of information by reading. Demerits—Probably a little tendency to a vagrant style of reading but this was converted to advantage (reasons given)." (spelling of "Mr. Neasy at Northwalk" uncertain) **B** *"Was admirably taught between the ages of thirteen and sixteen and a half years, to reason, use my own mind, and depend on myself. Was taught to acquire large masses of information by reading. There was a little tendency to a vagrant style of reading, but this was probably neutralized by other influences."*

JOHN MARSHALL

1. Qualities: Health: *"Very good, with slight exceptions. F. Very good except occasional gout. f. Apparently delicate, but all her family were long lived."* **D** Mental Peculiarities: "(Abstract). Steadiness; impartial; social affections strong; religious bias towards natural religion strong, as distinguished from dogma of any kind. Very curious about facts; love of present pursuit is very strong. F. An excellent *accountant,* and had had large and responsible business always in hand. M. A large brained, strongly opinioned woman; devoted to family matters. Y.S.B. General mental activity among all of them; all are strong and healthy. f.F. Very active certainly in their mental work and habits; sociable. f.M. More slow and cautious as a family, but

---

[66] In Galton's notebook on "Special Peculiarities" only two chemists are listed as having claimed high mechanical aptitude: Buckton and Lawes; since Buckton's reply is known, the remaining chemist quoted in *English Men of Science* on mechanical aptitude must be Lawes.

[67] Lubbock lived near Charles Darwin, who taught him a great deal of natural history.

[68] In 1835 Main was appointed chief assistant to Airy at the Royal Observatory

steady, circumspect, persevering and strong minded." [J] **Religious Bias:** *"Religious bias towards natural theology strong, as distinguished from dogma of any kind."*

NEVIL STORY-MASKELYNE

1. **Qualities: Health:** "Fair health with care. I reckon one fifth of my life has been lost in transient sickness—stronger now than when a boy—some sciatica. f. Good till later years." [D] **Mental Peculiarities:** "Remarkable dissimilarity in feature, form, and colour, of all my brothers and sisters and myself. There is very little family likeness among us. [It is not easy to extract further.]" [J]

2. **Origin of Taste for Science:** " '*I was always apt to observe stones closely with regard to their qualities'* [*but the scientific taste for geology was not developed till after manhood.*] [69]" (identification uncertain)   .

JAMES C. MAXWELL

1. **Qualities: Health:** "Often laid up before aet. 19, never since; never had a headache. F. Never had a headache; want of circulation latterly, d. aet. 64. f. Excellent health till aet. 41, she died the year after." [D] **Mental Peculiarities:** "Fond of mathematical instruments and delighted with the forms of regular figures and curves of all sorts. Strong mechanical power; extremely small practical business. Strongly affected by music when a child, could not tell whether it was pleasant or painful, but rather the latter; never forget melodies or the words belonging to them and these run through the mind at all times and not merely when the tunes are in fashion; can play on no instrument and never received instruction in music. Great continuity and steadiness; gratitude and resentment weak; στοργη [70] pretty strong; not gregarious; thoughts occupied more with things than with persons, social affections limited in range; given to theological ideas and not reticent about them; . . . constructiveness of imagination; foresight. F. Very great mechanical talent, and good business habits; strongly affected by music; of a mathematical turn of mind. Very steady; much partisanship; very great public spirit; extraordinary constructiveness of imagination; foresight. M. Guided by religious thought and very independent of the exhortations of acquaintances, clerical or lay. Religion was a forbidden subject in her father's family as his mother was a Roman Catholic and all discussion was avoided religiously." [J] **Business:** "Very low business." [A] *"Only two of my correspondents speak of being deficient in business capacities. Both these are physicists"* **Steadi-**

ness: "Great steadiness." [I] *"Great continuity and steadiness."* **Religious Bias:** *"Given to theological ideas, and not reticent about them."*

2. **Origin of Taste for Science:** *"I always regarded mathematics as the method of obtaining the best shapes and dimensions of things; and this meant not only the most useful and economical, but chiefly the most harmonious and the most beautiful. . . . I was taken to see . . . ,[71] and so, with the help of 'Brewster's Optics' and a glazier's diamond, I worked at polarization of light, cutting crystals, tempering glass, etc. I should naturally have become an advocate by profession, with scientific proclivities, but the existence of exclusively scientific men, and in particular of . . . ,[72] convinced my father and myself that a profession was not necessary to a useful life."*

3. **Education:** "Schools 10–16. Edinburgh University. Cambridge, 2nd wrangler. At home till 10, exclusively. Observation—encouraged by my father's knowledge of animals & by my relatives' practice of drawing. Merits—Living in a house where there were many interests & going to a day school where the boys were seen at a time when they were fresh and active. I had thus two worlds to balance against one another. On the whole I had the greatest freedom possible to a boy." [B] *Education included French, German, logic, natural philosophy, chemistry, besides mathematics. I lived in a house where I saw many people whose interests were of various kinds, and I went to a day-school where I mixed with the boys only when they were fresh and active. Thus I had two outer worlds to balance against each other. On the whole, I had, I think, the greatest degree of freedom possible to a boy."*

CHARLES MERRIFIELD

1. **Qualities: Health:** "Generally good. F. Early life good, later asthma badly—in late life fair, still alive, aet 85! f. In early life fair; in middle and later life, delicate." [D] **Energy:** "Physical—'A strong walker and oarsman. Endurance rather than great strength'; Mental—'Have always worked long hours and very fast. Can *write* more rapidly than any man I ever met, 32 folios of 72 words = 2160 an hour.' " [C] *"A strong walker and oarsman; can write more rapidly than any man I ever met (thirty folios of seventy-two words, equal to 2,160 words, an hour). In mind—Have always worked long hours and very fast. Father—Remarkable energy and endurance, notwithstanding asthma: very hard-working as a . . . . Mother—Physically weak, but has had a large family; has done a great deal of original as well as of steady work."* **Mental Peculiarities:** "Practical business habits; a ready speaker; ready at contrivance, mechanical and others; a mathematician, very persistent; cool and strong nerve; public spirit; curious about facts and caring little for argument;

---

[69] "He was persuaded to abandon the law for science in 1847 by Benjamin Brodie the younger . . . , and in 1850 was invited to deliver lectures on mineralogy at Oxford." *Dictionary of National Biography.*

[70] "στοργη" = "affection, fondness, tenderness, love"—A. Kyriakides, *A Greek-English Dictionary* (Nicosia, Cyprus, 1892); pertains especially to love of family.

[71] Mr. Nicol.

[72] Professor Forbes?

sceptical. F. Practical business habits; ready speaker and good dialectician. Very persistent; cool and strong nerve; strong partisan but public spirited. M. Practical business habits; musician and artist. Y.S.B. (Alike in form and physique but) great diversities of temper and dispositions. f.F. (as above in form and physique) violent and obstinate in temper. f.M. (also as above, in physique)." J Business: "Practical business habits." A Independence: "Yes, different political views to those of my own family; F. Yes, an advocate of Catholic views at a time when it was rare." E (reading of "rare" uncertain) *"Maintenance by my father of religious and political creeds at a time when these creeds were unpopular and often disqualifying."* (identification uncertain). Scholarliness: "Yes." H Steadiness: "Very persistent. Cool and strong nerve." I

2. Origin of Taste for Science: "I cannot say." F *"I cannot say whether they were innate. I was always brought up in a half-scientific, half-literary atmosphere, and was a fair mathematician as a boy, as well as a fair classic and linguist. My tastes were not determined by after-events, but my advocations were rather determined by my scientific habits."* (identification uncertain).

3. Education: "School 8–14. 0 honors. Pretty much self-taught. Was taught languages at home by private tutors. Observation—yes, encouraged in every way to use my eyes." B *"Pretty much self-taught, but encouraged to use my eyes, wits, and independent thought."*

JOHN MIERS

1. Qualities: Health: "Excellent during my long life. F. No returns. f. No returns." D Mental Peculiarities: "Mechanism and decided business activity. F. A steady man of business." J

2. Origin of Taste for Science: *"As a youth, I followed, of my own free-will, mineralogy, chemistry, anatomy, and mechanics, but chiefly chemistry. My tastes were certainly not hereditary. They were directed to botany purely through accidental circumstances [which led to a prolonged residence in an imperfectly civilized country].*[73] *I examined its plants, then wholly unknown to Europeans, but was at that time wholly ignorant of the very elements of botany. Was subsequently encouraged by . . . [eminent botanists of the day]* [74]; *went to and from England and made extensive collections. My wife actively assisted me in my botanical and other scientific pursuits, and to her advice and assistance I owe much of my success in life."*

WILLIAM H. MILLER

3. Education: "School, 15–19. Cambridge, 5th wrangler. Up to 15 was taught at home. Not condu-

[73] Chile and Argentina. "In 1818 Lord Cochrane invited Miers to join him in developing the copper and other mineral resources of Chile. . . ." *Dictionary of National Biography.*
[74] Robert Brown and John Lindley.

cive to observation? Not to health—rather reverse. Demerits—"Want of system." B *"Want of system."*

GEORGE MIVART

1. Qualities: Health: "Generally good especially since aet. 30. F. Apparently good in youth, but apoplexy aet. 50, d. aet. 74. f. Generally excellent, d. aet 64." D Mental Peculiarities: "(Insufficient returns) F. Very social and fond of pleasure and though irascible, very kind, doing benevolent things secretly. M. Very kind and charitable, often in a private way." J

2. Origin of Taste for Science: "Thoroughly innate." F *"Thoroughly innate. I had no regular instruction, and can think of no event which especially helped to develop it. Bones and shells were attractive to me before I could consider them with any apparent profit, and books of natural history were my delight. I had a fair zoological collection by the time I was fifteen. My father had no scientific knowledge; nevertheless, he encouraged me in all my tastes, giving me money freely for books and specimens, against the advice of friends; but he was indulgent generally, and not in the scientific direction only."*

3. Education: "Nothing especially to praise or blame. My father was indulgent in every way and allowed me plenty of money to buy specimens and books." B

WILLIAM NEWMARCH

1. Qualities: Health: "Excellent, but not very robust, but free from trouble—some ailment, though much improved since aet. of 30. F.. No returns. f. No returns." D Mental Peculiarities: "Have a special aptitude for public speaking and greatly delight in its exercise. Have a special aptitude for business and the organization of business arrangements. Like music but have no technical skill in it. Considerable steadiness of purpose and character, no very strong feelings of partisanship; great curiosity about facts of all sorts and about new discoveries; no belief in the marvellous, considerable constructiveness of imagination. F. Had a fair faculty of public speaking. Impulsive. M. Impulsive." J

ALFRED NEWTON

1. Qualities: Health: "Excellent, but hay fever. F. Excellent, but severe hay fever. f. Good." D *"Excellent, but hay-fever; father, excellent, but severe hay fever."* Energy: "Physical—'Endurance of rough travel when my companions suffered much discomfort'; Mental—'Can commonly work 12 to 14 hours a day without any particular exhaustion.'" O *"Sufficiently patient of ordinary fatigue, cold, hunger, to enable me to enjoy traveling in unfrequented countries when my companions suffered much discomfort. In mind—Can commonly work from twelve to fourteen hours a day*

without any remarkable amount of exhaustion. Father —*Capable of enduring fatigue."* **Mental Peculiarities:** "[Strong natural history.] Y.S.B. My youngest brother has my tastes for natural history equally developed in him, though circumstances have hindered him from indulging it to the same extent." [J]

2. **Origin of Taste for Science:** "Yes. I cannot recollect the time when I was not fond of animals and of knowing all I could respecting them." [F] *"I cannot recollect the time when I was not fond of animals, and of knowing all I could learn about them. Living in the country, I had abundant opportunities for indulging my taste, though, of course, I was not allowed to keep half the number of 'pets' I should have liked. The example of my father and elder brothers, who were all pretty firm to field-sports, was also followed by me, and from field-sports to field natural history is but a step. I obtained, by a piece of sheer good luck, the traveling fellowship of . . . ;* [75] *it was tenable for nine years, and its income was sufficient to keep me during that time without being obliged to enter any profession. Though circumstances subsequently interfered with my using this assistance to the most advantage, in gratifying my taste for natural history, it was enormously furthered thereby."*

3. **Education:** "Schools from 15–17. Cambridge. Taught by mother & sister till 15, then, a regular tutor. Conducive to observation on the whole, but variously; to health, eminently. Merits—My master was a man of scientific and generally liberal turn of mind. Demerits—The very mistaken way in which languages were taught, but I was idle." [B] *"My master (between the ages of fifteen and seventeen) was a man of scientific and generally liberal turn of mind." "The very mistaken way in which languages, as it now seems to me, especially Latin and Greek, were taught."*

SHERARD OSBORN

1. **Qualities: Health:** "Always very healthy. F. Always very healthy. f. Ditto, until Indian climate broke it down." [D] **Energy:** " 'Abundant energy, activity, and restlessness. Fond of adventure & travel'; Mental—'Much mental activity.' " [C] *"Very active in business, preferring walking to the compulsory driving; occupied fourteen or fifteen hours a day without distress; restlessness kept under conscious restraint; longing for adventurous travel, but hindered. Of mind— I doubt whether any one in my profession has done more work, if I may reckon the total work done in . . . , etc., etc.; and I worked nearly as hard while a student. Father—As a young man, an active cricketer and volunteer officer. A very earnest, active man in business,*

---

[75] Magdalene College, Cambridge. "From 1854 until 1863 he held the Drury traveling fellowship, making use of the endowment in the study of ornithology, a subject to which he had been attached from boyhood." *Dictionary of National Biography.*

heavily engaged in it from the age of eighteen. Besides, he took an active part in town affairs and the management of many associations. Mother—*A good walker, very active in the management of her house. Although she had a very large family, and took most diligent care of them, she was always at work, collecting all manner of things, arranging, describing, corresponding, painting, copying; she was never idle."* **Mental Peculiarities:** "Practical and business habits. Impulsive; of strong feelings and social affections.; a lover of the new; imaginative; with considerable foresight. F. A steady energetic soldier. M. Strong affections and deeply impressed wtih the advantage of giving her children a good and sound education." [J] **Business:** "Practical and business habits." [A] **Independence:** "Independent in all; F. Independent in all; f. not independent" [E] **Memory:** "Good." [G] **Scholarliness:** "Not studious, but receptive." [H] **Steadiness:** "Impulsive."

2. **Origin of Taste for Science:** "Innate." [F]

3. **Education:** "Schools, 8–13. Taught to read, write & speak badly. Self-taught—small amount of mathematics; taught on board HMS Excellent. Conducive to observation—so far as it went. Merits—none especially. Demerits—essentially defective, no competition or supervision." [B] *"Essentially defective; no competition nor supervision."*

RICHARD OWEN

1. **Qualities: Energy:** "Made an Alpine pedestrian tour aet. 60. Grouse and deer stalking at 67." [C] *" 'At the age of sixty made a tour, chiefly pedestrian, of four weeks in the Alps; ascended Cima di Jazi; crossed St. Théodule pass, walking sometimes thirty miles a day; aged sixty-seven, grouse-shooting and deer stalking. Walk six miles daily to present date.' Of mind—see list and dates of works and papers [an enormous amount of work]."* **Mental Peculiarities:** "Take a part in a 'Beethoven trio' with violincello. M. An accomplished organist and pianist." [J] **Independence:** ["Refers to his works. Power of God, etc.]." [E] **Memory:** "Habitually lectured without notes." [G] *"Although I can speak for an hour or two from a few notes, I could not repeat correctly a few sentences from memory. Father—Remarkable for good verbal memory; could repeat pages of poetry and speeches without mistake— a striking contrast to my own memory."* (identification uncertain)

2. **Origin of Taste for Science:** "Innate and derived from my mother." [F] *"Homology innate, and derived from my mother. I trace the origin of my interest in science decidedly to my mother's observations in our childhood rambles, on the plants and animals we saw. She told me that crabs were 'sea-spiders,' and periwinkles (Littorinae) 'sea-snails.' I feel sure she had never read 'De Maillet!' . . . ."*

3. **Education:** "School, 6–16. Edinburgh University, 2 years. During one year (aged seventeen) I

resided with my uncle [by marriage] and learnt there more of the dead languages than in all my previous school time. Conducive to observation—yes. To health—yes, during the ten years of school I lived at home in a hilly country, active out of door life. Merits—Great proportion of time left free to do as I liked. Demerits—Elements of natural science omitted, nothing taught of the nature of the world around us." B *"The great proportion of time left free to do as I liked, unwatched and uncontrolled." "Elements of natural science omitted; nothing taught of the nature of the world around us." "During one year (aged seventeen) I resided and studied with my uncle [by marriage], and learned there more of the dead languages than in all my schooltime."*

JAMES PAGET

1. **Qualities: Health:** "Good but often interrupted by acute illness. F. Extremely good, never ill but one day. D. of old age aet. 82. f. Very good, interrupted only by the birth and care of 17 children." D **Energy:** "Physical—'Very active, 14 or 15 hours a day without distress!'" Mental—'I doubt if any one in my profession has done more work!'" O **Mental Peculiarities:** "Practical business habits; music; considerable facility in public speaking, very small impulsiveness; strong partisanship and social affections. F. Strong business habits; steadiness. M. Strong partisanship and social affections; much religious bias of thought, love of the new and marvellous and curiosity about facts. Y.S.B. Of a numerous family nearly all have on various occasions shown great mental activity; there is a marked uniformity of character among them. F. and M. ordinary rather. One of M's sisters very energetic." J **Business:** "Practical business." A **Scholarliness:** "Always very studious." H **Steadiness:** "Steadiness." I

2. **Origin of Taste for Science:** "In a great degree innate." F *"Innate in a great degree. I trace the origin of my interest in science (1) to my mother's mental activity and love of collecting and arranging, and my father's constant encouragement of my pursuits; (2) to the friendship of [three eminent botanists], by whom I was chiefly induced to study botany; (3) to my profession, the choice of which was in some measure determined by my taste for collecting and studying."*

3. **Education:** "No school nor university; aet. 16 was apprenticed to a surgeon; aet. 20, Bartholomew's Hospital—many prizes. No regular teaching at home but study was greatly encouraged the habits of the house were orderly and laborious. To observation—neither. To health—fairly—in nothing." B

WILLIAM K. PARKER

1. **Qualities: Health:** "Not good originally & broke down. F. Great energy but highly nervous & broke down. f. Good till birth of first child, not after." D

Energy: "Mental—'My mind is frightfully active. Little education. Bad health, apprenticed to a trade yet successful in practice and science.'" O *"[This is a case of extraordinary mental activity, as shown by evidence which I do not feel justified in quoting. It was rewarded by success, notwithstanding serious impediments in boyhood.]* [76] Father—*A most energetic man; all for practical pursuits.* Mother—*An unusually strong mind, and steady love and search for knowledge."* **Mental Peculiarities:** "If I have any special talent, it is for drawing; none of my relations ever made the slightest sketch. I was busy drawing and colouring at 6 years of age; my . . . has thriven from this very thing, I do all things at a white heat, but never tire of a pursuit, but I learned to bridle my impulsive temperament at 15. My feelings are quick but I have no tendency whatever to partisanship. I have been the more biased towards religion in that my father and maternal grandfather *lived* it and did not prate about it. I am personally only a combination of these two men. I must have facts, and my mind is as fresh as ever in the love of the marvellous and beautiful in nature. F. In a large sphere he would have been a thorough benefactor. (Most striking instances of his conduct are give. Ed.) M. Was my father's help-meet in all sorts of kindnesses; he considered her to be the making of his family and especially of me. Y.S.B. We are a very unlike lot; my brothers' tastes are like my own, but the fire smoulders. f.M. Strong love of knowledge, though very busy people (Ample, confirmatory traits and instances are given. Ed.)" J **Scholarliness:** "Yes, but work hastily." H **Steadiness:** "I do all things at a white heat, but never tire." I *"I do all things at a white heat, but never tire of the pursuit."* **Religious Bias:** *"I have been the more biased toward religion, in that my father and maternal grandfather lived it, and did not prate about it. I am personally only a combination of these two men in this respect. (Please . . . take the sense of what I have written, and not the words.)"*

2. **Origin of Taste for Science:** "Innate." F *"I should say innate. I caught at all scraps of lessons for self-improvement. My soon-developed enthusiasm must have been derived from my mother's family. As to whether they were largely developed by events occurring after manhood, I think not. All I can say is, that neither profession nor marriage nor sickness has been able to affect them."*

3. **Education:** " 'Self-taught in drawing and caught at all the scraps of lessons for self-improvement. Conducive to observation—such as it was. To health—neither way, most of my time in the fields with my father's servants.' Merits—excited a quenchless search

[76] W. K. Parker "was second son of Thomas Parker, a yeoman farmer. . . . His early education at the parish school was obtained in the intervals of work on the farm. . . . On finally leaving shcool, he was apprenticed to a druggist at Stamford, under conditions which involved fifteen hours' work a day." *Dictionary of National Biography.*

for more of what I got so little. It merely started me & then I went alone." [B]

EDMUND A. PARKES

1. **Qualities: Health:** "Pretty good in youth not good for many years apparently due to dissecting wound. F. Always feeble; scrofulous ophthalmia in youth. f. Good." [D] **Energy:** "Yes, mental fatigue is a sensation not known." [C] *"At times, great fatigue has been gone through in connection with my profession. In mind—A good deal of continued power of brain-work; mental fatigue is a sensation not known.* Father *—Very energetic.* In mind—*remarkably so. Having been ruined in early life, he articled himself to a solicitor when he was thirty-five years of age; procured good practice, and wrote [a small technical book] on law.* Mother—*Loved to go through much fatigue."* **Mental Peculiarities:** "Nothing very special; certainly there is a love of pursuit of scientific facts and desire to advance my profession in public usefulness; I think also there is a steadiness in pursuit and not much impulsiveness. F. A very good business man. Y.S.B. A great deal of industry I think, in all. f.F. Many have been good businessmen, that is tradesmen, bankers, lawyers, etc., for 5 or 6 generations. f.M. My mother had 12 brothers and sisters, all of whom appear to have had unusual ability." [J] **Independence:** "F. Very independent; f. A very independent wise judging woman." [E] **Memory:** "Yes." [G]

2. **Origin of Taste for Science:** "I perceive no evidence (one way or the other)." [F] *"I perceive no evidence of their being innate [?hereditary], unless I derived any tendency from my mother, who was at one time much with her great-uncle [... the founder of one of our great industries],* [77] *and greatly interested in his pursuits. She worked a good deal at chemistry, and was well acquainted with many of the processes in pottery. I belonged to an industrious family, and saw every one working. The attraction I have for chemistry (which is a strong one, only my profession has never allowed me to follow it very closely) arose from being sent to work, at the age of fifteen, in a chemical laboratory."*

3. **Education:** "Schools, 5 years. A good deal of self-teaching. Rather unfavorable to health. Merits—none in particular. Demerits—careless superficial teaching." [B] *"Careless and superficial reading."*

JOHN PHILLIPS

1. **Qualities: Health:** "Never ill for more than two or three days except with neuralgia; no surgical operations except inoculation, drawing of one tooth, and cutting of corns. F. Not known. f. Believes to have been good." [D] **Mental Peculiarities:** "Inventiveness—max-

imum thermometer. . . . Mathematics chiefly turned to natural philosophy problems. Desire to know whatever came naturally before me and to employ on every subject all that could be known. On this account zoology, botany, geology, mineralogy, meteorology, astronomy have all in turn and often in repetition been my delight. . . . I think a large range of information and much public speaking have been my recommendations to success." [J]

LYON PLAYFAIR

1. **Qualities: Health:** "Never very robust, broke down from brain work aet. 43–44; recovered after. F. Good health generally. f. Never very robust." [D] **Energy:** "Mental—'Can work for long periods and get through various works without mixing them.'" [C] **Mental Peculiarities:** "Strong practical business habits which have forced upon him incessant work in public capacities; love of public work and little care for its bearing on personal interests. F. Excellent man of business which led to high employment in his profession; love of work and singularly disinterested in character. M. Strong curiosity for facts; powerful social affections." [J] **Business:** "Strong practical business habits." [A] **Independence:** "Tolerably; f. Independent in thought." [E] **Memory:** "Considerable. Can learn a speech of an hour's duration by merely reading it 3 times." [G] *" 'After reading over a lecture or speech of an hour's duration, three times, can recollect nearly the words as written for 8 or 10 days.' [I am informed verbally by this correspondent, that he is obliged to abstain from writing out his addresses, etc., beforehand, otherwise he has found the memory of what he wrote to be so strong and exacting as to make it difficult to him to deviate from it and accommodate his language to the current temper of his audience.]* Mother—*Excellent memory."* **Scholarliness:** "More receptivity than studiousness." [H]

2. **Origin of Taste for Science:** "Scarcely innate." [F] *"Scarcely innate. I ascribe the origin of my scientific interests chiefly to being sent as a pupil to an eminent man of science, Prof. . . .* [78] *subsequently I was a good deal abstracted from scientific pursuits by an early and lasting friendship with . . . , who directed my thoughts to public work."*

3. **Education:** Schools, 8–14. University, 15–21—five of these are below. A few class prizes, nothing else. Merits—Coming in contact with persons of every rank and sitting in the same forms with sons of tradesmen and ploughmen as well as gentlemen. Demerits—too great changes of system—5 universities, 3 Scotch, London, and Giessen." [B] *"A variety of subjects and attention to details. Coming in contact with persons of every rank [in Scotland], and sitting on the same*

---

[77] See discussion of Wedgwood ancestry in *English Men of Science*, pp. 62–64.

[78] Probably Professor Thomas Graham, then at Andersonian Institute, Glasgow.

form with the sons of tradesmen and ploughmen, as well as of gentlemen." "Too great changes in system, having been educated at five universities (three of which were Scotch, one London, and one in Germany)."

WILLIAM POLE

1. **Qualities: Health:** "Strong. F. Always good. f. Fair general health, not strong." D **Energy:** "Mental —'Considerable judging from the number of things I have been able to learn & do.'" C *"I should say considerable, judging by the number of things I have been able to learn and do since adult age.'"* **Mental Peculiarities:** "Music by nature considerable; by education high; [79] mathematics with special reference to their useful practical application, high; practical business habits high. I attribute all the knowledge I have acquired and any success I have had chiefly to three qualities, all of which I believe I inherited. 1. Independence of judgement which prompted me to learn for myself what I wanted to know. 2. Earnestness, determination and perseverance in acquiring such knowledge, often under difficulties and in the face of routine business occupation. 3. A businesslike, practical, logical way of looking at things, which enabled me to direct attention to the important and relevant, neglecting the unimportant and irrelevant points in what I had to study and do. F. Remarkable earnestness in anything he seriously undertook; devoted the last few years of life entirely, zealously, and laboriously to religious and charitable objects. M. Sound common sense—practical views of things." J **Business:** "Practical business habits." A *"I attribute all the knowledge I have acquired, and any success I may have had, chiefly to three qualities, all of which I believe I inherited. First, independence of judgment which prompted me to learn for myself what I wanted to know. Secondly, earnestness, determination, and perseverance in acquiring such knowledge, often under difficulties, and in the face of routine business occupation; and thirdly, a businesslike, practical, logical way of looking at things, which enabled me to direct attention to the important and relevant, neglecting the unimportant and irrelevant points in what I had to study and do."* **Independence:** "Very independent indeed and will not trust to any body in authority; f. Independent in character." E **Scholarliness:** "Varied acquirements." H **Steadiness:** "Earnestness and determination." I

2. **Origin of Taste for Science:** *"Innate, I think, as regards certain qualities of mind, which led me, under the pressure of circumstances, to direct my attention to certain things in a certain way, namely, (1) independence of judgment; (2) earnestness of purpose; (3) a practical. clear-headed, common-sense, logical way of viewing things."*

3. **Education:** "Schools to about 16. Self-taught in

---

[79] Pole was a musician as well as an engineer, and he was concerned as well with the physics of music.

all but the usual education of common English schools. Demerits—omission of Greek & the violin—almost everything else may be learned in after life." B

JOSEPH PRESTWICH

1. **Qualities: Health:** "Very fair. F. Excellent till within a few months of death aet. 78. f. Excellent but undermined by bleeding as was then much the practice after confinement." D **Energy:** "Physical—'Considerable. Worked all day at business & ½ or ¾ the night at science. From Saturday till Monday morning walked 40 or 50 miles geologizing'; Mental—'Worked all day at business and found wonderful relief in science at which I worked during many hours at night.'" C *"Used to work all day at business, and one-half or three-quarters of the night at science. From Saturday afternoons to Monday mornings would walk forty to fifty miles [in pursuit of a branch of natural history]. Could work hard at business all day (and a very anxious business), and at evening and night would work hard at [two branches of science]. Found a wonderful relief in science. Father—Energetic in traveling; great energy in business."* **Mental Peculiarities:** "Sciences of observation. Impulsiveness and perseverance; strong feelings; not much religious bias except in a boundless admiration of nature; as a boy, much love of the new and marvellous and curiosity about facts; strong love of my own pursuits; foresight and I believe disinterestedness. F. Music, drawing, impulsiveness, strong feelings, social affections; love of the new and curiosity about facts. M. Singular disinterestedness. f.M. Great shyness in some, especially my mother; a very painful hereditary complaint." J **Independence:** "Chalked out my own studies; F. Marked. Held Jacobite opinions when it was not safe to shew them." E *"It was marked in my father; he held Jacobite opinions, when it was not very safe to hold them."* **Scholarliness:** "Irregular, never tired of certain branches of study." H **Steadiness:** "Impulsiveness & perseverance. Strong love of my own pursuits." I **Religious Bias:** *"Not much religious bias except in a boundless admiration of nature."*

2. **Origin of Taste for Science:** "Yes, I can well recollect as a boy of 6 etc. etc." F *"As well as I can recollect, they were innate. I remember, as a boy of six, seeing a spring in Lavender Hill; not being satisfied at the explanation, and determining to work it out for myself. I believe that I should have devoted myself to chemistry and physics, but that I was started, as a youth of nineteen, to travel ten months out of the twelve on business, and so continued for twenty years. This led to my visiting all Great Britain, and to great opportunities for geologizing, and determined me to that study. I worked hard at business all day (a very anxious business), and at evening and night would work hard at chemistry and geology. I found a wonderful relief in science."*

3. **Education:** "Various schools, 5–16, one of them

in Paris. 14–16, Dr. Valpy reading—usual prizes. 16–18, University College, London—2nd in chemistry. Self-taught entirely in geology & mostly in mathematics and natural philosophy. Conducive to observation—not at all. To health—fairly so. Merits—if any, its variety. Sufficently grounded in many subjects to avoid errors. Knowledge of French and drawing. Demerits —Want of system. Neglect of many subjects for the attainment of one or two. Not pushing in mathematics to a useful end." [B] *"Sufficient ground-work in many subjects to avoid error." "Neglect of many subjects for the attainment of one or two; not pushing mathematics to a useful end." "Want of system."*

ANDREW C. RAMSAY

1. **Qualities: Health:** "Delicate in early life, strong as a man and capable of enduring great fatigue. F. Good up to about 58, then paralysis. f. Always good, slipped out of life aet. 84." [D] **Mental Peculiarities:** "Musical; good enough man of business; understands accounts, etc. [A modest statement for a man in high official position! Ed.] F. The preceeding I should think would apply to my father. M. Excellent business habits as she proved when left a widow with 4 children." [J]

HENRY ROSCOE

1. **Qualities: Health:** "Good. F. Good; died early aet. 35. f. Good." [D] **Energy:** "Physical—'Restless energy till last year, when health has been less satisfactory." [C]
2. **Origin of Taste for Science:** "Yes, I was always observing and inquiring." [F] *"I was always observing and inquiring, and this disposition was never checked nor ridiculed in my childhood. My taste for chemistry dates from the lectures I attended as a boy, and to the permission to carry on little experiments at home in a room set apart for the purpose. I was encouraged in my tastes at home. Subsequent determining events were my residing abroad, and my mother making a home for me there."*
3. **Education:** "8–16, School. 16–24 University College and Heidelberg. Exhibitor for chemistry & a summa cum laude—Heidelberg. To observation—yes, always observing and inquiring and the disposition never was checked nor ridiculed in my childhood. To health—not specially. Merits—Not being tied down to the old courses of classics & mathematics. To encouragement by my mother—to Prof. Williamson and to Bunsen." [B] *"I was fortunate in obtaining at school (between the ages of eight and sixteen) an insight into the phenomena of Nature—a subject entirely ignored at that time in almost all schools. My peculiar bent for experiment was encouraged at home by my mother, and there were peculiar merits in my training under Profs. . . . at . . . ,[80] and especially in Germany under*

. . . ." [81] *"Not tied down to old courses of classics and mathematics."*

LAURENCE PARSONS, EARL OF ROSSE

1. **Qualities: Health:** "Not a very robust constitution, but not subject to illness. F. No returns. f. No returns." [D] **Mental Peculiarities:** "Fond of machinery and mechanism of all kinds; perhaps not sufficiently orderly and methodical to become a good man of business. Fond of music though it was not made a part of my edution nor was the taste for it specially encouraged. No very great quickness for pure mathematics nor for the higher branches of either 'pure' or 'applied.' Not impulsive; not imaginative. F. Fond of mechanism and mechanical engineering, also of military engineering. Fond of music but in early life laid it aside in favor of scientific pursuits especially astronomy, and also of public business. Not impulsive. Y.S.B. All of rather retiring disposition perhaps increased by *home* education. f.F. Perhaps shyness and retiring disposition." [J]
2. **Origin of Taste for Science:** *"Primarily derived [both by inheritance and education] from my father."* (identification uncertain) [82]

EDWARD SABINE

1. **Qualities: Health:** "Generally very good. F. Generally very good. f. Generally very good." [D]
2. **Origin of Taste for Science:** *"Largely determined by my service in north-polar and equatorial expeditions."* [83] (identification uncertain)

ROBERT SALISBURY

1. **Qualities: Health:** "Bad in youth, fair in middle age. F. Good through life. f. Bad: died aet. 40." [D] **Independence:** "F. Extremely obstinate." [E]
3. **Education:** "Schools & Eton to 15. Nominally classics, practically nothing. Oxford—feeble in health and learnt very little. So far as I have learnt anything, self-taught. Restrictive of observation? The reverse of conducive to health. Demerits—Disregard of health, limitation of subjects practically to classics, very isolated at home. Brought up in an idle class & never realized the necessity of labour in acquirement." [B] *"Limitation of subjects practically to classics." "Brought up in an idle class, and never realized the necessity of labor in acquirement."*

J. BURDON-SANDERSON

1. **Qualities: Health:** "Always moderate. F. Good. f. Good." [D] **Mental Peculiarities:** "Aptitude for

---

[80] Williamson at London.

[81] Bunsen at Heidelberg.

[82] Rosse's father was the astronomer William Parsons, third earl of Rosse.

[83] In 1818 Sabine accompanied J. Ross on his expedition in search of the North-West Passage, and later to the tropics of Africa and America.

mechanism, probably attributable to the desultory method of my education and my having lived in youth in a mining country. Apt at acquiring modern languages; fond of mathematics but knowledge of the subject defective from my imperfect education (Great steadiness of charatcer and inquisitiveness of mind. Never satisfied without following a subject to its foundations. Great dislike for superficiality. Thorough in every thing he undertakes; the most trifling and insignificant thing must be done in the best possible way. Utterly disinterested, *communicated by his wife*.) F. Possessed remarkable powers of writing and speaking English." [J] **Mechanics:** "*Aptitude for mechanism.*"

PHILIP SCLATER

1. **Qualities: Health:** "Excellent. F. Excellent. f. Generally good." [D] **Energy:** "Am a bold rider with hounds; member of Alpine Club and not easily tired." [O] "*I am a hard rider with hounds, fond of mountaineering, and not easily tired. Father—An active man all his life, riding every day, and always about, although over eighty.*" **Mental Peculiarities:** "I believe I may say that my organ of order is highly developed. Of my collection of some 7000 birdskins every one is always in its place, ticketed with name etc. all by my own hands. I spend much time (perhaps too much) in putting things straight." [J] **Business:** "Order highly developed. Of my collection of 7000 birds' skins, every one is always in its place, ticketed with name, etc., all by my own hand. I spend much time, perhaps too much, in putting things straight." [A] "*I believe I may say that my organ of order is highly developed. Of my collection of some seven thousand birds' skins, every one is always in its place, ticketed with name, etc., all by my own hand. I spend much time, perhaps too much, in putting things straight.*"

2. **Origin of Taste for Science:** "I believe so; I began collecting birds and studying them before I went to school and without any inducement." [F] "*I cannot trace the origin of my interest in geology. I believe it to have been innate. I began collecting birds and studying them before I went to school, and without any inducement. I was always told by my relations that my scientific pursuits would stand in my way, but adhered to them notwithstanding. They were not at all determined by events occurring after I reached manhood; they simply increased as I grew older.*"

3. **Education:** "Schools—Winchester 12–16. Oxford. 1st class mathematics, a fellowship. Not conducive to observation." [B]

ROBERT SCOTT

1. **Qualities: Health:** "Very good when young and only suffering from effects of sedentary pursuits. F. Good throughout life, died aet. 72. f. Weak, died of senile phthisis aet. 72." [D] **Energy:** "Physical—'Cap-

tain of Trinity College, Dublin, eleven for 5 years— great activity at cricket, football. Used to row a great deal in heavy boats at sea.' " [O] "*Great activity at cricket and football up to age twenty-five. Captain of . . .*[84] *eleven for five years; used to row a great deal in heavy boats.*" **Mental Peculiarities:** "Strong business habits. Partisanship, curiosity about facts; love of pursuit; patience in working and results, all strongly developed. Nearly absolute absence of originality; dislike of metaphysics." [J] **Business:** "Strong business habits." [A] **Independence:** "Very slight." [E] **Memory:** "Very retentive but not exactly accurate." [G] "*Very retentive but not exactly accurate.*" **Scholarliness:** "Very fond of study; i.e. of work." [H] **Steadiness:** "Patience in working out results strongly developed." [I]

2. **Origin of Taste for Science:** "I had an innate wish for miscellaneous information." [F] "*I cannot say, except that I had an innate wish for miscellaneous information. My interest in science arose from the chance circumstance of my choosing civil engineering as a profession, and having spare time, when studying at . . . , which I devoted to . . . . My scientific tastes were subsequently determined by my not having any profession, except civil engineering, which I never followed.*"

3. **Education:** "Rugby 2–18. Dublin, Berlin, Munich. Many honors at Dublin. Tutor at home to the aet. 12. To observation—not especially except in regards geological work. To health—yes. Merits—It was very general & extensive. Demerits—Too much cramming for exams, too much isolated. I formed no friendships being youngest son and educated at home." [B] "*Too much cramming for examinations. Too much isolated, being the youngest son, and educated at home.*"

WILLIAM SIEMENS

1. **Qualities: Health:** "Good condition, active, restless, easily fatigued but persevering. F. Good organization. f. Healthy when young but subject to severe disorders in after life having had 14 children." [D] **Energy:** "Physical—'Active habits, restless disposition; easily fatigued but persevering; naturally adventurous'; Mental—'Capable of sustained effort in pursuit of interests.' " [O] **Mental Peculiarities:** "Mechanism and physical principles; facility in concluding amicable arrangements in business matters and in carrying them through in a conciliatory spirit. Irresistible desire to realize objects in applied science when I am convinced they have been pitched too high and have led to fruitless expenditure of means and energy, although the principles have been proved. Domestic habits, with love of the new but not of the marvellous." [J] **Independence:** "Independent; F. Independent with high moral standard." [E] **Memory:** "Very retentive and can easily forget useless things." [G] "*Never kept a diary; clear remembrance of events in childhood with their dates in every year from the age of six onward. Solve*

---

[84] Trinity College, Dublin.

problems better out-of-doors than in the study. Can forget useless knowledge, such as formulae, rules, gossip, etc., very fast." **Scholarliness:** "Not naturally studious, but receptive and imaginative." [H] **Steadiness:** "Sanguine and determined in objects of importance." [I]

3. **Education:** "One year at Göttingen—then had to to go to business." [B][85]

ARCHIBALD SMITH

1. **Qualities: Health:** "Very good in youth and early manhood, weakened in middle life by intense work; subsequently broken by overwork. F. Good but not very strong. Had malaria in early manhood which left his health delicate. f. Fair." [D] **Mental Peculiarities:** "(Extract) Mathematics [very strongly], mechanism. Practical business habits, great steadiness of purpose; too balanced a mind to be a paritisan; warmly attached to friends; great interest in Biblical studies; delight and interest in new scientific discoveries. Caution in forming opinions. Not observant in ordinary life. Simple tastes; fond of animals. [Absolutely true; his moral and intellectual character always seemed to me *one* in its truth and accuracy—*Communication* from an intimate friend.] F. General abilities in many directions but not very special. Very impractical, for instance, he could never be got to take an interest in the mines in his own property because they do not belong to the geological formation he had made his study. Steadiness of purpose; warm social affections; great interest in religious subjects; great simplicity of character; truthfulness and high sense of honor; fond of animals. M. Musical; practical ability; warm family affections; absorbing interest in religious subjects, great truthfulness, very silent and reserved; great refinement, fond of animals. Y.S.B. Some sisters very musical, some very deficient in the science of numbers; great fondness for animals. f.F. Very retentive memories, especially for people, genealogy, &c. One uncle knew the Navy List by heart, another the Army List. Great fondness for animals. f.M. Musical—a talent for cutting intagtios and modelling [transmitted]." [J]

3. **Origin of Taste for Science:** "[*The following is an extract from biographical notes kindly communicated to me of the late Archibald Smith.*] *'Yachting would give an interest to all nautical matters, and the intimacy of his father with . . . gave a bias toward magnetism. In a letter to one of his sisters (no date,? about 1838), he says: ". . . told me he was going to write directions for ships, finding and allowing for the error caused by the local attraction of ships. So, for my own amusement and partly to help him, I wrote a set of in-*

structions and gave them to him." His mind was thus turned to the subject. I think it was natural to him to inquire into the reason of things. Fond of figures when a boy.'"

PIAZZI SMYTH

1. **Qualities: Health:** "Good. F. Good. f. Good." [D] **Independence:** "Preference for whatever is not the fashion, not popular, not rich, not very able to help itself, and yet with qualities unworthily overlooked or unjustly oppressed; F. Yes. f. Very independent character." [E] *"Preference for whatever is not the fashion, not popular, not rich, not very able to help itself, yet with qualities unworthily overlooked or unjustly oppressed."*

2. **Origin of Taste for Science:** *"Ocean-voyaging in beginning of life. Solitary observing for years in an observatory,*[86] *placed in a country verging on a desert, but under southern skies rich in stars unknown to the ancients, and not appreciated by the moderns."*

WARINGTON SMYTH

1. **Qualities: Health:** "Good. F. Good to within 2 or 3 years of death aet. 77. f. Good to within a short time of death aet. 85." [D] **Energy:** "Physical—'Energy of body and power of enduring rough travel in small means in Asia Minor, etc. Have rowed myself in a skiff 105 miles in 21 hours. Rowed 2 years in a Cambridge University crew.'" [O] *"Considerable energy and power of enduring fatigue; rough traveling on small means in . . . [partly-civilized countries]. Have rowed myself in a skiff one hundred and five miles in twenty-one hours while undergraduate at . . . ;*[87] *rowed in every race during my stay at the university; rowed two years in the university crew [Oxford and Cambridge races]. Father—[Many examples of his energy in his life]. Of mind—considerable, compiling and writing on a great variety of subjects, while at the same time carrying on a system of . . . observations,*[88] *and for years together. Mother—Energy of mind very similar to that of my father; joining nightly in . . . observations daily in writing or drawing . . . ."* **Mental Peculiarities:** "A decided good and innate taste for music. Love of pursuit of one's own branch of science. . . ." [F] **Unwearied industry, public spirit and disinterestedness; much delight in imaginative work; strong feelings and partisanship; religious bias of thought strong, yet with scepticism as to many commonly accepted matters. M. Steadiness; love of the new and marvellous in science; curiosity about facts and love of pursuit strong." [J] **Independence:** "Yes, often find myself at a truce with popular opinions; F. Yes. f. Maintained very free views on social and religious topics even when being

---

[85] "After being educated in the polytechnic school of Magdeburg and the University of Göttingen, he visited England at the age of nineteen in the hope of introducing a process in electroplating invented by himself and his brother Werner." *Dictionary of National Biography.*

[86] Royal Observatory, Cape of Good Hope.

[87] Cambridge.

[88] Smyth's father was W. H. Smyth, who was engaged in the Admiralty Survey of the Mediterranean.

surrounded by very different opinions." [E] **Memory:** "Very moderate." [G]

2. **Origin of Taste for Science:** "Yes (decidedly)." [F] *"I believe I may say innate, to a very considerable extent, not remembering that any definite steps were taken to inculcate science. I was indebted in a high degree to collections made by my father and mother, in . . . ,*[89] *and to early familiarity with charts of those seas, and conversations on matters pertaining thereto; afterward, to going to Germany and finding in the mining officers a body of men receiving a regular scientific education; lastly, to a great extent by going for a winter to . . . [in Germany] and by conversations with . . . and . . . ."*

3. **Education:** "Westminster, Bedford, Cambridge. Almost entirely self-taught as regards natural science. Not conducive to observation. Yes, to health. Demerits at school—too much for memory, nothing for thought. I could now wish I had gone through at the University of good course of chemistry & physics as a preparation for the other branches but the main obstacle was lack of time." [B] *"I could now wish that I had gone through, at the unversity, a good course of chemistry and physics, as a preparation for the other branches; but the main obstacle was lack of time." "Too much for memory; nothing for thought."*

HERBERT SPENCER

1. **Qualities: Health:** "Rather feeble as a child, grew up stronger, aet. 35 broke down. F. Originally strong. Health gave way from overwork aet. 33, remained nervous for many years but by care became healthy as he got towards 60. Strong when old. f. Rather feeble when young, subsequently of medium strength." [D] **Mental Peculiarities:** "I think my own powers are very evenly balanced. I have inherited various things and have tolerable power of manipulating; take pencil portraits; modelled a bust; sing, especially part music; pretty good in mathematics so far as I went; methodic in work and business. But I am below par in linguistic faculty, so far as it is shown in learning languages, which I always hated. I do not care for history as it is ordinarily written and have a contempt for biography (Quaere, as to autobiography? Ed.) Decidedly steady rather than impulsive; moved to continuous action by a large aim. Decided love of the novel. No unusual curiosity about facts for themselves, but only with a view to conclusions. Great constructiveness of imagination. Very fond of castle-building as a boy. Always tending towards the ideal, detecting faults and looking for remedies. Unusual faculty of seeing kinship between things not apparently related. Strong sense of public duty leading in many cases to disregard of self-interest. The emotional traits I regard as having led to success, are strong self-esteem and disregard of authority and intolerance of

wrong whether in acts or opinions. Of the intellectual traits I may say that after constructiveness of imagination which I think I have in a very unusual degree, the most important peculiarity is the even balance of the analytic and synthetic tendencies,—also an uncommon charatceristic. F. My father's intellectual power and my own were nearly akin. He had however more artistic skill and greater manipulative dexterity. He might have been a first rate portrait painter and a good sculptor. He was inventive but not in wholly novel ways; he devised by far the best system of short hand which exists. Not at all a man of business. Wanting in judgment as to proportion between means and ends. Much of what I have said of myself applies to my father; but there were these differences. He was moved to activity by small ends; large ones paralysed him. More curious about facts for themselves than I am. Extremely strong sense of public duty. M. Orderly and somewhat methodic. Not at all impulsive; very uniform and balanced in conduct. Very religious in the genuine, unsuperstitious way; philanthropic in more than average degree. Great love of the new and marvellous. Extremely disinterested in the domestic relations. f.F. All the brothers extremely independent; tending to eccentricity. All given to discussion, often ending in disputes. All intellectually active; all but one (a lawyer) conscientious and having a high sense of public duty. My father the most marked of them." [J]

WILLIAM SPOTTISWOODE

1. **Qualities: Health:** "Good. F. Good. f. Good." [D] **Energy:** "[Enormous quantity of work, FG]." [C] **Mental Peculiarities:** "No returns [high business powers, FG] F. Mechanism; practical business habits. M. Was a good musician [member of a family having high business capacities,[90] F.G.]" [J] **Business:** "Practical busniess habits." [A]

2. **Origin of Taste for Science:** *"My interest in mathematics began at . . . [university], and was mainly due to the energy and encouragement of my tutor . . . ; but Prof. . . . first inspired me with the sense of the magnificence of mathematics."* (identification uncertain)

3. **Education:** "Eton, Harrow, Oxford—1st class mathematics." [B]

JOHN STENHOUSE

1. **Qualities: Health:** "Generally paralysis aet. 28. Always delicate, infl. of eye. F. Health delicate, apoplexy and inflam. eyes; died aet. 41. f. Delicate but lived to 75." [D]

3. **Education:** "Schools, 6–13. Glasgow University & Giessen. Have been taught French & German to some extent. Merits—Was very well grounded in arith-

---

[89] The Mediterranean?

[90] Spottiswoode's father was a partner in Eyre & Spottiswoode, printers.

metic at school and learned natural history under James D. Forbes, who was a very good teacher. Demerits— Want of modern language & of chemistry."[B] *"Was very well grounded in arithmetic at school." "Want of the modern languages and of chemistry."*

BALFOUR STEWART

1. **Qualities: Health:** "Generally good. F. Good. f. Died of consumption after a premature confinement."[D] **Energy:** "Physical—'Have great vitality, restless'; Mental—'No power of great amount of work.'"[C] **Mental Peculiarities:** "Great love of the new and marvellous and also great curiosity about facts; and great power of abstraction. Was in early life very fond of pure mathematics, aet. 16 or 17 I conceived independently Lagrange's calculus of fractions."[F] **Independence:** "Very early a lover of justice and truth. Left aet. 12 a school where I received injustice from the master. Entertained at an early age independent religious views which led to frequent disputes at home."[F] *Left at the age of twelve [that is ran away from], a school where I had received injustice from the master.* **Memory:** "Good for poetry."[G] **Religious Bias:** *"Entertained at an early age independent views regarding the resurrection and salvation of the heathen, which led to frequent disputes."*

2. **Origin of Taste for Science:** "Yes, I remember incidents before I could write."[F] *"[Yes.] 'I remember [incidents which proved an innate taste quoted at length] before I could write, [but] I believe the origin of my pursuit of physical science was when I attended the natural philosophy class at . . . .[91] I was intended for business, but, conceiving a distate for it, I left it, and attached myself to science.'"*

WILLIAM STOKES

1. **Qualities: Health:** "Good in early life; subject to headache. F. Good; subject to headache. f. Good."[D] *"Good in early life, subject to headache; father, good, subject to headache."* **Energy:** "Physical—'Great energy without restlessness'; Mental—'Great mental energy.'"* **Mental Peculiarities:** "Practical business habits; great love for music, love of painting and art but does not practise them. Steady, with strong feelings but not partisan; warm social affections; slow to make new friendships; of a religious bias of thought; good story teller and inventor of stories for children; great foresight and public spirit with utter disinterestedness. F. No practical business habits though hardworking; great love of music, fond of poetry which he wrote and art though not practising any branch of it. A man of honour of a high and pure nature, the very sole of honour. He was singularly beloved by a large circle of friends, of a deep religious bias and strong

opinions which he would have borne out to the death. Inclined to sacrifice everything, family interest and personal welfare to public good and conscience. M. Very practical, lover of music, played the violin; fond of poetry and art though not producing them. Of warm affections, agreeable in society but not a woman of the world; practical and independent in spirit; she was anxious for the welfare of her family more than caring for public interests. Y.S.B. High artistic power in some of its members."[J] **Business:** *"Business habits."*[A] **Independence:** "Yes, very; F. Yes; f. Yes."[F] **Memory:** "Very remarkable."[G] **Scholarliness:** "Yes."[H] **Steadiness:** "Steady with strong feelings."[I] **Religious Bias:** *"Of a religious bias of thought."*

2. **Origin of Taste for Science:** "Yes; they appear to be inherited (from my father)."[F] *"They appear to have been inherited. My interest in science arose from the example of my father[92] and the fact of my being for a year the assistant and close companion of Prof. . . . ,[93] of . . . ,[94] at whose side I visited the poor in the lanes of . . . ,[95] day and night. First began to work and concentrate energies to one branch at the age of twenty-one, when appointed . . . .[96]"*

3. **Education:** "Schools, none. Edinburgh University, medicine. Glasgow in chemistry. To observation —yes, especially. To health—yes, most of my boyhood in open air. Merits—Great number of studies connected with nature. Demerits—Want of system and consecutive study."[B] *"The merits of my education consisted in the great number of studies connected with nature; but there was a want of system and of consecutive study."*

RICHARD STRACHEY

1. **Qualities: Health:** "Good average. F. Good average. f. Good average."[D] **Mental Peculiarities:** "Mathematical tendency; power of organization in official business combined with inventive faculty in an administrative sense. Capacity for appreciation of numerical data,—accounting, finance, statistics, and deriving principles from them. F. Great practical business habits, great love of work. His sense of duty, his love of justice, his affection for his children, which was more that of a father's than a mother's, were such as I have never seen in the same proportion in any but one other man, and by him they were not surpassed. Y.S.B. All the males are intelligent above average, and decidedly peculiar though it is difficult to say in what way, the present tendencies are distinctly discernable in the preceding generation and may possibly have existed fur-

---

[91] Probably the natural philosophy class of James D. Forbes at Edinburgh.

[92] Stokes' father was Whitley Stokes, Regius Professor of medicine at Dublin.

[93] Robert James Graves?

[94] Dublin.

[95] "In the great epidemic of typhus in Dublin in 1826 his exertions in the treatment of the poor were conspicuous." *Dictionary of National Biography.*

[96] Physician to the Meath Hospital.

ther back. Besides the intellectual qualities of tendency towards acquisition of precise knowledge in detail, capacity for business, active political interest, with an absence of piety, want of sporting interests, general inaptitude for squire life and indisposition to personal exaltation there is a certain eccentricity of manner . . . combined with most decided opinions. f.M. My mothers and her two sisters were highly intelligent." J Religious Bias: *"The negative tendencies of my family may be absence of piety."*[97]

2. Origin of Taste for Science: *"A mathematical tendency, I think, led me first toward . . . inquiry, to which I have been faithful ever since. Professional duties and civil engineering kept up a disposition to appreciate the material constituents of the world, and led, through surveying, in the direction of physical geography. The distinct origin of my desire to place myself among scientific students was the wonderful impression produced on me by the aspect of nature, as seen in the . . . ,*[98] *combined with what I may call the accident of my having been allowed to explore a part of it in an official capacity. Having thus made rather large botanical and geological collections, I came to England with them, and, while employed in arranging and distributing them, picked up a certain rather irregular and unsystematic scientific education, in the company of . . .*[99] *and others. Forced back into professional life special scientific inquiry has not been possible; but I have had opportunities*[100] *of aiding the progress of science, which I have endeavored to make the best of."*

ALLEN THOMSON

1. Qualities: Health: "Weak as a child, never very robust, rheumatism and catarrh. F. Strong in youth, asthma and rheumatism in middle life, otherwise healthy. f. Robust and healthy during greater part of life, latterly catarrh and bronchial affections." D **Energy:** "Physical—'Active'; Mental—'Was capable of 16 or more hours work, daily, in earlier life.' " O **Mental Peculiarities:** "A decided turn for mechanical pursuits both in arrangement and construction; practical business habits; no very strongly marked mental peculiarities. F. A natural turn for mechanical pursuits, not however cultivated to any extent beyond what might be expected in an operating surgeon; good business habits. Great love of literature and philosophy and science, with power of acquisition of varied knowledge, arrangement and generalisation; strong will; perseverance; warm

[97] In the notebook on "Special Peculiarities" Galton's page discussing religion singles out De La Rue and Strachey as having an "absence of piety."
[98] Nepal.
[99] In Nepal "he made the acquaintance of Major F. Madden, under whose guidance he studied botany and geology, making explorations into the Himalaya ranges west of Nepal for scientifiic purposes." *Dictionary of National Biography.*
[100] Strachey was at the time of Galton's questionnaire on the council of the Royal Society; he was also involved on committees concerning meterology.

friendships and strong partisanships; views of religion philosophical; character impulsive rather than carefully prudent; unselfish and disinterested, public spirited; charitable in feeling and action. He raised himself entirely by his own exertions and under great difficulties from a mechanical employment to the head of his profession of medicine. M. Very practical in her habits of management; a good ear and natural taste for music. Love of literary and historical rather than of scientific knowledge; steady perseverance; matter of fact rather than imaginative; affection very strong; charitable and disinterested. f.M. Most of her family were lively in disposition, decided and independent in their judgements; careful and steady in their habits with strong feelings and affections; some shewed a decided turn for mechanical construction." J **Business:** "Practical business." A **Independence:** "Yes, but not anxious to obtrude my opinions; F. Though strictly educated formed independent religious social and political judgements; f. considerable." E **Memory:** "Not good." G **Scholarliness:** "Considerable range but not capable of very deep thought." H **Steadiness:** "Impulsive in philosophy & religion." I **Mechanics:** *"A decided turn for mechanical pursuits, both in arrangement and construction."*

2. Origin of Taste for Science: "I believe in nearly equal degree the mixed result of natural bias and education." F *"They have been, I believe, nearly in an equal degree the mixed result of a natural bias and education, and were determined by professional study, when a love of scientific knowledge for its own sake first took possession of my mind."*

3. Education: "Schools, 8–14. Glasgow, 14–21. To observation—rather restrictive. To health—not adverse. Merits—My private education at home was much the more valuable. Demerits—Those common to most of the Scotch schools of the period, which were directed to the cultivation of literary tastes only and not adapted therefore to a variety of intellects." B *"My private education at home was much the more valuable." "School-work directed to the cultivation of literary tastes only, and therefore not adapted to a variety of intellects."*

JOHN VOELCKER

1. Qualities: Health: "Delicate at an early age, never perfect health, but became stronger especially after aet. 29. F. Delicate. f. Good and robust." D **Independence:** "Brought up a Lutheran, joined the Independents & became Baptist by conviction." E

2. Origin of Taste for Science: "From an early age I had an innate taste for all branches of science." F *"From an early age I had an innate taste for all branches of natural science. As a boy, I made large collections of dried plants, minerals, beetles, butterflies, stuffed birds, etc. At . . . I studied without regard to future profession for two years, and only took up chemistry as a special study on my third year's residence there."*

3. **Education**: Schools. Göttingen & Utrecht—a scholarship at the former . Self-taught between 18 & 22 before I could find the means of going to Göttingen. To observation—specially so. To health. Merits— Early training in habits of observation & good teaching. As a boy I made large collections of various objects in natural history." B

THOMAS WATSON

1. **Qualities: Health**: "Good. F. Good. f. Good." D **Energy**: "Indolent [but capable of work.] C *"When a boy of thirteen I walked forty-eight miles in one day, fifty the next and about twenty the third; when grown up, my powers were ordinary, certainly not above the average.* In mind—*Naturally indolent; disinclined to work unless with a large object.* [*N.B.—I insert this moderate statement because my correspondent adheres to it verbally, and gives facts and reasons which I cannot controvert; nevertheless, if energy is to be measured by work actually accomplished, and if my correspondent's work be compared with that of other men, the estimate of his energy would be prodigiously increased.*[10]] Father—*When a young man he and two brothers walked sixty miles in one day. Much mental energy; ready for all purposes. When old he was astonished at the amount of work in . . . he did when young.* Mother— *Ordinary, both bodily and mental."* **Steadiness**: "Indolent." H

2. **Origin of Taste for Science**: "No." F *I cannot say that I had naturally a turn for any pursuit in particular. My addiction to medicine was purely the result of accident. I never gave a thought to physic as a subject of study until I was twenty-seven years old.*[102] " (identification uncertain)

3. **Education**: "Schools, Cambridge—10th wrangler." B

ALEXANDER W. WILLIAMSON

1. **Qualities: Health**: "Very delicate in childhood, always well when not overworked. F. Strong constitution but much dyspepsia. f. Exceedingly healthy but not especially strong." D **Mental Peculiarities**: "No returns. M. Strong sense of duty; great simplicity and rectitude of character; singularly truthful habit of mind; steadfast love of justice; strong religious feeling with entire absence of bigotry or sectarian feeling. Y.S.B. Strength of purpose (of will) in self and sister. f.M. Similar general charatceristics to those of my mother." J

WILLIAM C. WILLIAMSON

1. **Qualities: Health**: "Very excellent, but threatened some years ago through overwork. F. Has been

good, but lumbago and sciatica in middle lite. Now 80, has bronchitis but is vigorous comparatively. f. Very good till died of bronchitis aet. 70." D **Energy**: "Physical—'Above the average; active work an absolute necessity for my being'; Mental—'My work must answer this.' " C *"Active bodily work an absolute necessity of my being; without it my epigastrium would gnaw itself into fiddle-strings. In mind—My scientific works must answer this question [they are very considerable].* Father—*Decidedly active and energetic; used to go out fossil-hunting when it was too late to follow his occupation*[103] *[which involved out-of-door work, lasting all day and fatiguing to the muscles].* Mother—*Very industrious.*" **Mental Peculiarities**: "Practical business habits, a decided quality; talent for drawing and manual work generally; decided steadiness in definite pursuits; social affections fervent; decided religious bias of thought; intense love of pursuits. F. Mechanical talent shewn in considerable skill in taxidermy. Decided steadiness in business pursuits. He is still as interested in his horticulture as when a youth; attending entirely to his plants when not prevented by illness. M. Business habits decided (facts given in evidence), steady in her business pursuits and in her painting, social affections most intense and womanly; one of the most unselfish of women." J **Business**: "Decided business habits." A **Independence**: Left ch. Engl. for Wesleyanism; F. Never gave his adhesion to any church." E **Scholarliness**: "Considerable." M **Steadiness**: "Decided steadiness in definite pursuit." I **Religious Bias**: "*Religious bias of thought decided.*"

2. **Origin of Taste for Science**: "I cannot judge, but I doubt their innate character." F *"My father's example influenced me so early that I have no means of judging, but I doubt much their innate character. Their origin was due primarily, beyond all probability of doubt, to my father's influence and example. They were not influenced by subsequent events, but the tastes once planted rather determined the events. My medical profession caused me to suspend my scientific pursuits for some years; but the accidental perusal of . . . brought me back again to the study of the . . . , and all the rest followed in due time."*

3. **Education**: "Schools. London, University College, of honors never a scrap, good bad or indifferent. Chiefly self-education, especially in science. To observation—the lack of school discipline left me free. To health—yes, because it was lax and interrupted and interesting. Merits—being born with some hydrocephalic tendencies the defects of my school probably saved my life. Demerits—Neglect of mathematics. Mental training overlooked in the mere acquisition of routine school and knowledge." B *"Neglect of mathematics; too much reliance on mere work of memory. Mental training overlooked in the mere acquisition of routine."*

---

[101] Watson's medical writings included many hundreds of pages.
[102] At the age of twenty-seven, Watson began the study of medicine at St. Bartholomew's Hospital.

[103] Gardner and first curator of the Scarborough Museum.

PHILIP J. YORKE

1. **Qualities: Health:** "Fair health. F. Not robust, died aet. 47. f. Good health." [D]
3. **Education:** "Harrow, army." [B]

2. IDENTIFICATIONS

The following pages indicate the identifications which have been made of the replies quoted by Galton in *English Men of Science.* The page numbers cited are those which appear in the first edition of Galton's book. The numbers in bold type correspond to those in the following index table, whereas the other numbers are the same as those used by Galton in the text of *English Men of Science.*

I. *Chapter II ("Qualities") pp. 74–143*

A. "Energy Much Above the Average——Forty (*sic*) Cases," pp. 78–98.
  #1 Forbes **1**
  #2 Hooker **2**
  #3 Günther **3**
  #4 James Bateman **4**
  #5 —— **5**
  #6 P. M. Grey-Egerton **6**
  #7 George Bentham **7**
  #8 A. Cayley **8**
  #9 Warington Smyth **9**
  #10 S. Osborn **10**
  #11 A. Harcourt **11**
  #12 —— **12**
  #13 D. Ansted **13**
  #14 R. Hill **14**
  #15 R. Owen **15**
  #16 ——**16**
  #17 G. H. Parkes **17**
  #18 W. C. Williamson **18**
  #19 T. Cobbold **19**
  #20 —— **20**
  #21 T. Watson **21**
  #22 J. F. Bateman **22**
  #23 W. Fairbairn **23**
  #24 Merrifield **24**
  #25 P. L. Sclater **25**
  #26 C. Darwin **26**
  #27 W. Fergusson **27**
  #28 —— **28**
  #29 —— **29**
  #30 R. H. Scott **30**
  #31 S. Haughton **31**
  #32 A. Newton **32**
  #33 W. K. Parker **33**
  #34 —— **34**
  #35 —— **35**
  #36 —— **36**
  #37 J. Prestwich **37**
  #38 T. Andrews **38**
  #39 —— **39**
  #40 —— **40**
  #41 W. Pole **41**
  #42 —— **42**
B. "Cases of Energy Below the Average—Two Cases," pp. 97–98.
  #1 W. S. Jevons **43**
  #2 F. Currey **44**
C. "Health," pp. 99–102.
  #1 (p. 100) ——**1**

#2 (p. 100) Phillips **2**
#1 (p. 101) Newton **3**
#2 (p. 101) Stokes **4**
#3 (p. 101) Ansted **5**
D. "Perseverance," pp. 103–104.
  #1 R. Main **1**
  #2 T. Hirst **2**
  #3 W. De La Rue **3**
  #4 J. S. Maxwell **4**
  #5 J. Lawes **5**
  #6 —— **6**
  #7 H. J. Carter **7**
  #8 W. K. Parker **8**
  #9 C. Darwin **9**
  #10 W. Armstrong **10**
  #11 W. Fairbairn **11**
E. "Practical Business Habits," pp. 104–107.
  #1 C. Darwin **1**
  #2 P. L. Sclater **2**
  #3 T. H. Huxley **3**
  #4 A. Farre **4**
Two physicists—J. C. Maxwell and W. Grove **5 & 6**
Enumeration of three qualities—W. Pole **7**
F. "Memory," pp. 107–121.
  1. "Good verbal memory, as for prose and poetry," pp. 109–111.
    #1 —— **1**
    #2 L. Playfair **2**
    #3 J. Evans **3**
    #4 —— **4**
    #5 —— **5**
    #6 —— **6**
  2. "Good memory for facts and figures," pp. 111–113.
    #1 —— **7**
    #2 R. Hill **8**
    #3 —— **9**
    #4 F. Evans **10**
    #5 —— **11**
    #6 J. F. Bateman **12**
    #7 —— **13**
    #8 W. Siemens **14**
    #9 D. Forbes? **15**
  3. "Good memory for form," pp. 113–116.
    #1 J. Ball **16**
    #2 A. Günther **17**
    #3 J. D. Hooker **18**
    #4 G. Back **19**
    #5 —— **20**
    #6 W. Fergusson **21**
    #7 —— **22**
  4. "Good memory for names in natural history," pp. 116–117.
    #1 T. Cobbold **23**
    #2 G. Bentham **24**
    #3 P. Grey-Egerton **25**
    #4 —— **26**
  5. "Good memory, no particulars given," p. 117.
    #1 —— **27**
    #2 Fairbairn? **28**
    #3 —— **29**
    #4 R. H. Scott **30**
    #5 —— **31**
    #6 W. Fox **32**
  6. "Fitful and peculiar memory," pp. 118–120.
    #1 ——**33**
    #2 R. Owen? **34**
    #3 V. Harcourt **35**
    #4 —— **36**
    #5 C. Darwin **37**
    #6 W. De La Rue **38**
  7. "Bad Memory," pp. 120–121.

#1 —— 39
#2 —— 40
#3 R. Carrington 41
#4 W. Guy 42
#5 —— 43
#6 —— 44
#7 —— 45
G. "Independence of Character," pp. 121–124.
  1. Lines 5–14 from top, p. 122.
    #1 B. Stewart 1
    #2 —— 2
    #3 —— 3
    #4 P. Smyth 4
  2. Lines 1–7 from bottom, p. 122 and p. 123.
    #1 —— 5
    #2 R. Hill 6
    #3 —— 7
    #4 J. Prestwich 8
    #5 Merrifield? 9
H. "Mechanical Aptitudes," pp. 124–126.
  #1 Buckton 1
  #2 Lawes 2
  #3 J. Evans 3
  #4 W. Carpenter 4
  #5 Farre 5
  #6 Fergusson 6
  #7 Gray 7
  #8 Henslow 8
  #9 Huxley 9
  #10 and #11 Sanderson and —— 10 & 11
  #12 Thomson 12
  #13 Greg 13
  #14 Hill 14
  #15 Jevons 15
I. "Religious Bias," pp. 126–141.
  1. "Religious sentiments weak; accompanied with more or less Scepticism," p. 130.
    #1 —— 1
    #2 Strachey 2
    #3 De La Rue 3
    #4 —— 4
    #5 Prestwich 5
    #6 Darwin 6
    #7 —— 7
  2. "Intellectual Interest in Religious Topics," pp. 130–131.
    #1 Stewart 8
    #2 —— 9
    #3 Maxwell 10
    #4 Main 11
  3. "Dogmatic Interest," p. 131.
    James Bateman 12
  4. "Religious Bias," pp. 131–134.
    #1 Allman 13
    #2 Stokes 14
    #3 Buckton 15
    #4 Ball 16
    #5 J. H. Balfour 17
    #6 Parker 18
    #7 A. C. Williamson 19
    #8 —— 20
    #9 Gladstone 21
    #10 Haughton? 22
    #11 Cobbold 23
    #12 —— 24
  5. "Religious bias, with intellectual scepticism," pp. 134–136.
    #1 Frederick Evans 25
    #2 Forbes 26
    #3 Marshall 27

    #4 Carpenter 28
    #5 Huxley 29
  6. "Have no dread of inquiry," p. 136.
    #1 —— 30
    #2 —— 31
    #3 —— 32
  7. "Religion and science have different spheres," p. 136.
    #1 —— 33
  8. "Indifference to dogma," p. 137.
    #1 Cobbold 34
  9. "Liberality of early religious teaching," pp. 137–138.
    #1 —— 35
    #2 —— 36
    #3 —— 37
    #4 —— 38
    #5 —— 39
    #6 —— 40
    #7 —— 41
    #8 —— 42
    #9 —— 43
  10. "Have early abandoned creeds," pp. 138–139.
    #1 —— 44
    #2 —— 45
    #3 Darwin 46
  11. "The religious creed has had a good effect on freedom of research," pp. 139–140.
    #1 —— 47
    #2 —— 48
    #3 —— 49
    #4 —— 50
    #5 —— 51
    #6 Günther 52
    #7 —— 53
    #8 —— 54
  12. "Has had some deterrent effect," pp. 140–141.
    #1 —— 55
    #2 —— 56
    #3 —— 57
    #4 —— 58
    #5 —— 59
    #6 —— 60
    #7 —— 61
    #8 —— 62

II. Chapter II ("Origin of Taste for Science") pp. 144–234.
A. "Physics," pp. 149–157.
  #1 W. Lassell 1
  #2 S. Haughton 2
  #3 F. Evans 3
    #4 G. Foster 4
  #5 De La Rue 5
  #6 Andrews 6
  #7 B. Stewart 7
  #8 R. H. Scott 8
  #9 P. Smyth 9
  #10 —— 10
  #11 L. P. Rosse? 11
  #12 Grove 12?
  #13 R. Strachey 13
  #14 E. Sabine? 14
  #15 R. Main? 15
  #16 J. C. Maxwell 16
  #17 T. Hirst 17
  #18 A. Cayley 18
  #19. A. Smith 19
  #20 W. Spottiswoode? 20
B. "Chemistry," pp. 158–161.
  #1 A. Harcourt 21

#2 G. Buckton **22**
#3 H. Roscoe **23**
#4 J. H. Gladstone **24**
#5 J. Voelcker **25**
#6 —— **26**
#7 L. Playfair **27**
#8 —— **28**
#9 —— **29**
#10 —— **30**
#11 —— **31**

C. "Geology," pp. 161–164.
#1 J. Evans **32**
#2 T. R. Jones **33**
#3 D. Ansted **34**
#4 W. W. Smyth **35**
#5 P. M. Grey-Egerton **36**
#6 J. Prestwich **37**
#7 D. Forbes **38**
#8 M. Story-Maskelyne? **39**

D. "Biology," pp. 165–182.
1. "Zoological Subsection," pp. 165–176.
#1 J. E. Gray **40**
#2 C. Darwin **41**
#3 J. Lubbock **42**
#4 R. Owen **43**
#5 T. S. Cobbold **44**
#6 G. Mivart **45**
#7 G. J. Allman **46**
#8 W. K. Parker **47**
#9 A. Newton **48**
#10 —— **49**
#11 —— **50**
#12 —— **51**
#13 T. H. Huxley **52**
#14 W. Carpenter **53**
#15 —— **54**
#16 W. C. Williamson **55**
#17 W. Stokes **56**
#18 A. Thomson **57**
#19 A. L. C. Günther **58**
#20 P. L. Sclater **59**
#21 E. A. Parkes **60**
#22 J. G. Jeffreys? **61**
#23 —— **62**
#24 —— **63**

2. "Botanical Subsection," pp. 176–180.
#1 J. D. Hooker **64**
#2 J. Ball **65**
#3 J. Bateman **66**
#4 H. J. Carter **67**
#5 J. Miers **68**
#6 J. H. Balfour **69**
#7 J. S. Henslow **70**
#8 C. Babington **71**
#9 F. Currey **72**
#10 G. Bentham **73**

3. "Medical Subsection," pp. 180–182.
#1 W. Paget **74**
#2 —— **75**
#3 G. Humphry **76**
#4 W. Fergusson **77**
#5 —— **78**
#6 T. Watson **79**
#7 —— **80**

E. "Statistics," pp. 182–184.
#1 —— **81**
#2 Hatherly? **82**
#3 W. Guy **83**
#4 W. S. Jevons **84**
#5 R. Hill **85**
#6 J. Heywood **86**

F. "Mechanical Science," pp. 184–186.
#1 W. Armstrong **87**
#2 F. Jenkin **88**
#3 J. F. Bateman **89**
#4 W. Pole **90**
#5 Merrifield? **91**

G. "Analysis of Replies," pp. 186–222.
1. "Innate Tastes," pp. 186?–197.
a. "Physics and Mathematics": #1) Lassell, #2) Haughton, #3) Evans, #4) Foster, #5) De La Rue, #6) Andrews, #7) Stewart, #8) Scott, #11) Rosse?, #16) Maxwell, #17) Hirst, #18) Cayley.
b. "Chemistry": #1) Harcourt, #2) Buckton, #3) Roscoe, #4) J. H. Gladstone, #5) Voelcker.
c. "Geology": #1) J. Evans, #2) T. R. Jones, #3) Ansted, #4) W. Smyth, #5) Grey-Egerton, #6) Prestwich, #7) Forbes.
d. "Zoology": #1) J. E. Gray, #2) C. Darwin, #3) Lubbock, #4) Owen, #5) Cobbold, #6) Mivart, #7) Allman, #8) Parker, #9) Newton, #11) ——, #12) ——, #13) Huxley, #14) Carpenter, #15) ——, #17) W. Stokes, #18) Thomson, #19) Günther, #20) Sclater.
e. "Botany": #1) Hooker, #2) Ball, #3) Bateman, #4) Carter, #5) Miers, #6) J. H. Balfour, #7) Henslow, #8) Babington.
f. "Medical Science": #1) Paget, #2) ——.
g. "Statistics": #1) ——, #3) Guy, #4) Jevons.
h. "Mechanical Science": #1) Armstrong, #2) Jenkin.
2. "Instances of Tastes Being Decidedly Not Innate," pp. 191–192.
a. "Physics and Mathematics": #15) Main?
b. "Chemistry": #10) ——.
c. "Zoology": #16) W. C. Williamson, #22) Jeffreys?, #24) ——.
d. "Botany": #10) Bentham.
e. "Medical": #3) Humphry, #4) Fergusson, #6) Watson, #7) ——.
f. "Statistics": #2) Hatherly?
3. "Fortunate Accidents," pp. 198–199.
a. "Physics and Mathematics": #19 Smith.
b. "Chemistry": #1) Harcourt, #3) Roscoe, #9 ——.
c. "Geology": #2) Jones.
d. "Zoology": #9) Newton, #16) W. C. Williamson, #22) Jeffreys.
e. "Botany": #10) Bentham.
f. "Statistics": #4) Jevons.
g. "Mechanics": #2) Jenkin.
4. "Indirect Motives or Opportunities," pp. 199–201.
a. "Physics and Mathematics": #5) De La Rue, #8) R. H. Scott, #19) Smith.
b. "Chemistry": #6) ——.
c. "Geology": #1) J. Evans, #3) Ansted, #6) Prestwich.
d. "Zoology": #5) Cobbold, #9) Newton, #13) Huxley, #15) ——, #24) ——.
e. "Medical Science": #3) Humphry, #4) Fergusson.
f. "Mechanics": #3) J. F. Bateman, #4) ——.
5. "Professional Duties," pp. 202–205.
a. "Physics and Mathematics": #4) Foster, #9) P. Smyth, #13) Strachey, #14) Sabine?, #15) Main?
b. "Chemistry": #8) ——.
c. "Zoology": #1) Gray, #10) ——, #13) T. H. Huxley, #17) W. Stokes, #23) ——, #24) ——.
d. "Botany": #7) ——.
e. "Medical Science": #1) Paget, #2) ——, #3)

Humphry, #6) Watson, #7) ——.
f. "Statistics": #2) Hatherly?
g. "Mechanics": #2) Jenkin, #3) J. F. Bateman, #4) Pole.
6. "Encouragement at Home," pp. 205–211.
　a. "Physics and Mathematics": #10) ——, #11) Rosse?
　b. "Chemistry": #3) Roscoe, #4) Gladstone, #8) ——, #11) ——.
　c. "Geology": #1) J. Evans, #4) W. Smyth, #7) Forbes.
　d. "Zoology": #9) Newton, #15) ——, #21) Parkes, #1) Gray, #6) Mivart, #7) Allman, #11) ——, #16) W. C. Williamson, #17) W. Stokes, #19) Günther, #23) ——, #4) Owen, #8) Parker.
　e. "Botany": #2) Ball, #6) J. H. Balfour, #8) Babington, #10) J. Bateman.
　f. "Medical Science": #1) Paget.
　g. "Statistics": #5) Hill.
　h. "Mechanics": #5) Merrifield, #3) J. F. Bateman.
7. "The Influence and Encouragement of Friends," pp. 211–214.
　a. "Physics and Mathematics": #3) F. Evans, #6) Andrews, #13) Strachey, #16) Maxwell, #17) Hirst, #19) Smith.
　b. "Chemistry": #2) Buckton.
　c. "Geology": #2) Jones, #4) W. Smyth, #5) Grey-Egerton.
　d. "Zoology": #3) Lubbock, #10) ——, #23) ——, #24) ——.
　e. "Botany": #5) Miers, #9) Currey, #10) Bentham.
　f. "Medical Science": #1) Paget, #7) ——.
　g. "Statistics": #5) Hill.
　h. "Mechanical Science": #2) Jenkin.
8. "Influence and Encouragement of Tutors," pp. 215–218.
　a. "Physics and Mathematics": #7) Stewart, #10) ——, #20) Spottiswoode?
　b. "Chemistry": #7) Playfair.
　c. "Geology": #5) Grey-Egerton.
　d. "Zoology": #5) Cobbold, #17) W. Stokes, #24) Carter.
　e. "Botany": #4) Carter, #6) J. H. Balfour.
　f. "Medical Science": #4) Fergusson.
　g. "Statistics": #4) Jevons, #6) J. Heywood.
9. "Travel in Distant Parts," pp. 218–221.
　a. "Physics and Mathematics": #3) F. Evans, #9) P. Smyth, #13) Strachey, #14) Sabine?
　b. "Geology": #7) Forbes.
　c. "Zoology": #2) Darwin, #13) T. H. Huxley.
　d. "Botany": #5) Miers.
10. "Unclassed Residum," pp. 221–222.
　a. "Chemistry": #10) ——, #11) ——.
　b. "Geology": #8) Story-Maskelyne?

III. Chapter IV ("Education"), pp. 234–260.
　1. "Merits: Education Praised Throughout, or Nearly so—ten cases," pp. 238–241.
　　#1 Main 1
　　#2 Haughton 2
　　#3 Humphry 3
　　#4 Jevons 4
　　#5 Maxwell 5
　　#6 J. Evans 6
　　#7 Hill 7
　　#8 Roscoe 8
　　#9 F. Evans 9
　2. "Merits in Education: Variety of Subjects—Nine

Replies," pp. 242–243.
　　#1 Roscoe 10
　　#2 Newton 11
　　#3 Prestwich 12
　　#4 Jenkin 13
　　#5 Jevons 14
　　#6 Playfair 15
　　#7 & 8 Hooker is one 16 & 17
　　#9 Maxwell 18
　3. "Merits in Education: A Little Science at School—Three Replies," p. 243.
　　#1 Carrington 19
　　#2 Forbes 20
　　#3 Fox 21
　4. "Merits in Education: Simple Things Well Taught—Three Replies," pp. 243–244.
　　#1 F. Evans 22
　　#2 Stenhouse 23
　　#3 Carter 24
　5. "Merits in Education: Liberty and Leisure—Three Replies," p. 244.
　　#1 Maxwell 25
　　#2 —— 26
　　#3 Owen 27
　6. "Merits in Education: Home Teaching and Encouragement—Eight Replies," pp. 244–245.
　　#1 Roscoe 28
　　#2 —— 29
　　#3 Humphry 30
　　#4 Hill 31
　　#5 Owen 32
　　#6 Thomson 33
　　#7 —— 34
　　#8 Merrifield 35
　7. "Merits and Demerits in Education Balanced—Four Replies," pp. 245–246.
　　#1 Ball 36
　　#2 Stokes 37
　　#3 Back 38
　　#4 J. F. Bateman 39
　8. "Demerits: Narrow Education—Thirty-two Cases," pp. 246–251.
　　#1 Darwin 40
　　#2 Forbes 41
　　#3 W. Fergusson 42
　　#4 Guy 43
　　#5 Salisbury 44
　　#6 Lubbock 45
　　#7 Grey-Egerton 46
　　#8 Heywood 47
　　#9 Jevons 48
　　#10 Thomson 49
　　#11 Owen 50
　　#12 Harcourt 51
　　#13 Cobbold 52
　　#14 Stenhouse 53
　　#15 Jeffreys 54
　　#16 Hooker 55
　　#17 W. C. Williamson 56
　　#18 W. Smyth 57
　　#19 Jones 58
　　#20 —— 59
　　#21 Prestwich 60
　　#22 Carrington 61
　　#23 —— 62
　　#24 —— 63
　　#25 —— 64
　　#26 —— 65
　　#27 —— 66
　　#28 —— 67
　　#29 —— 68

#30 —— 69
#31 —— 70
#32 —— 71
9. "Demerits in Education: Want of System and Bad
   Teaching—Ten Cases," pp. 251–252.
   #1, #2, & #3 Miller, Prestwich, and Stokes 72, 73,
      74
   #4 Ansted 75
   #5 Babington 76
   #6 Osborn 77
   #7 Newton 78
   #8 W. Smyth 79
   #9 Fox 80
   #10 Parkes 81
10. "Demerits in Education: Unclassed—Four Cases,"
    p. 252.
    #1 Salisbury 82
    #2 Scott 83
    #3 Playfair 84
    #4 Gladstone 85

3. INDEX TABLE

In order to summarize the material on Victorian
scientists available through Galton's material, the following index table has been prepared.

The list of 192 names on this table comprises a complete list of all of those to whom Galton sent his questionnaire, as is determined by comparing his annotated
membership list of the Royal Society for November,
1872, with the various "Keys," and other material. All
of those who are known to have replied have been
given an asterisk, and for this latter group of 104 scientists some supplementary information (scientific field,
date elected to the Royal Society, father's occupation,
and higher education) has also been included where
easily ascertained from standard biographical sources.

An "x" in the table signifies that there exists manuscript material for a given scientist with respect to the
topic in the column heading, but that no identifications
have been made on this basis in the corresponding

pages of *English Men of Science*. An unbracketed
number the table signifies the existence of both manuscript material and an identification in the text. A
bracketed number indicates that an identification has
been made in the text, but from other sources (such as
the *Dictionary of National Biography*) rather than
from manuscripts. All numbers in the table correspond
to the bold type numbers for the respective pages in the
preceding identifications.

The columns of information on "stature," "head circumference," "political affiliation," and "religious affiliation" are taken from Galton's notebook labeled "Head
Circumference of Scientific Men." It should be noted
that in some cases Galton has entered anthropometrical
data in this notebook which was not derived from the
questionnaire but rather from his own measurements;
this fact explains his occasional notation "FG" alongside a measurment, as well as the fact that for some
scientists the anthropometrical measurement is the only
datum recorded. With respect to political and religious
persuasions it seems that Galton used "0" for "none."
However, when "0" is followed by some affiliation
(i.e., "0(C)"), it is probably to be interpreted as "nom-
inal" or "none . . . but leaning towards." The reader
may draw his own conclusions about the meaning of
"LC," "0(L-C)," which undoubtedly indicate subtle
political shadings. The lack of an entry on either political or religious affiliation cannot be interpreted to
mean "no affiliation."

The column labeled "Mental Peculiarities" indicates
whether or not a given scientist is referred to in the
"Extracts on Mental Peculiarities." Both the identifications of "Mechanical Aptitude" and "Religious Bias"
have been made from this source. The nine remaining
columns refer to topics which correspond to Galton's
nine "Keys." The section of the text entitled "Origin
of Taste for Science" corresponds to Galton's Key entitled "Innate Tastes."

Name	Scientific Field	Date F.R.S.	Father's Occupation	Education	Stature	Head Circumference	Political Affiliation	Religious Affiliation	Business Habits (pp. 104–107)	Education (pp. 234–260)	Energy (pp. 78–98)	Health (pp. 99–102)	Independence (pp. 121–123)	Orig. Taste Science (pp. 149–186)	Mechanical Aptitude (pp. 124–126)	Memory (pp. 107–121)	Mental Peculiarities	Religious Bias (pp. 126–191)	Scholarliness	Steadiness (pp. 103–104)
Frederick A. Abel (1827–1902)	Chem. Engin.	1860																		
Henry W. Acland (1815–1900)	Medicine	1847																		
James C. Adams (1819–1892)	Astronomy	1849																		
William G. Adams (1836–1915)	Physics	1872																		
George B. Airy (1801–1892)	Astronomy	1836																		
James Alderson* (1794–1882)	Medicine	1841	Physician	Cambridge	5' 9 1/2"		LC	ChE	X	X	X							X		
George J. Allman* (1812–1898)	Biology	1854		Dublin Oxford	5'10"				X	X	X			46				X	13	
Thomas Andrews* (1813–1885)	Chemistry	1849	Linen Merch	Glasgow, Paris Edinburgh	5' 5 3/4"	22 7/8"	O	ChE	X	X	38	X	X	6				X		X
David G. Anstad* (1814–1880)	Geology	1844		Cambridge	5'10"		C-L	ChE	X	75	13	5	X	34			X	X		X
James G. Apjohn (1796–1886)	Geology (Meteorology)	1853																		
Duke of Argyll (1826–1900)	Geology	1851																		
William Armstrong* (1810–1900)	Engineering	1846	Merchant	Bishop Auck. Gram. Sch.	5'10 1/2"	22 7/8"	L	ChE	X	X				87				X		10
James M. Arnott (1794–1885)	Medicine	1843																		
Charles Babington* (1808–1895)	Botany	1851	Physician	Cambridge	5'10 1/2"		L	ChE		·76		X		71						
George Back* (1796–1878)	Artic. Expl.	1847		Navy	5'9"	22 5/8"	LC	ChE		38	X	X	X			19	X			
John H. Balfour* (1808–1884)	Botany	1856	Army Surgeon	Edinburgh	5'9"	23						X		[69]				X	17	
Thomas S. Balfour* (1813–1891)	Medicine	1858	Merchant	Edinburgh	5'11 1/4"	22 7/8"						X								
John Ball* (1818–1889)	Botany	1868	Irish Judge	Cambridge	6' 3/4"	23 1/4"		KC	X	56	X	X	X	65		16		X	16	X
Henry C. Bastian* (1837–1915)	Medicine	1868		London	5'8 1/2"	21 1/2"						X						X		
James Bateman (1811–1897)	Horticulture	1838	Landowner	Oxford	5'9"					X		4	X	66		X	X		'12	X
John F. Bateman* (1810–1889)	Civil. Engin.	1860		Surveyor's Apprentice	5'11 3/4"	21 3/4"	C	Mor. ChE		39	22	X		89		12	X			
Lionel S. Beale (1828–1906)	Medicine	1857																		
Thomas S. Beck (1814–1877)	Medicine	1851																		
Thomas Bell (1792–1880)	Zoology	1828																		
John J. Bennett (1801–1876)	Mathematics	1841																		

Name (dates)	Scientific Field	Date F.R.S.	Father's Occupation	Education	Stature	Head Circumference	Political Affiliation	Religious Affiliation	Business Habits (pp. 104-107)	Education (pp. 234-260)	Energy (pp. 78-98)	Health (pp. 99-102)	Independence (pp. 121-124)	Orig. Taste Science (pp. 149-186)	Mechanical Aptitude (pp. 124-126)	Memory (pp. 107-121)	Mental Peculiarities	Religious Bias (pp. 126-141)	Scholarliness	Steadiness (pp. 103-104)
George Bentham* (1800-1884)	Botany	1862	Naval Architect	No formal Education	5'11 1/2"	22 1/2"	O(C)	ChE	X		7	X	X	73			24			
Rev. James Booth (1806-1878)	Mathematics	1846																		
William Bowman* (1816-1892)	Physiology	1841	Banker	Birmingham Hospital					X								X			
Benjamin C. Brodie (1817-1880)	Chemistry	1849				22 3/8"														
Charles Brooke* (1804-1879)	Medicine	1847	Mineralogist	Cambridge	5'8 1/2"	22 5/8"			X								X			
George Buckton* (1818-1905)	Chemistry (Entomology)	1857		Royal College of Chemistry	uni-form	22 3/4"	C	ChE	X	X			X			22	1	15	X	
George Burrows (1801-1887)	Medicine	1847																		
George Busk (1807-1886)	Biology	1850																		
William B. Carpenter* (1813-1885)	Physiology	1844			5'11 1/2"									53	4		X	28		
Richard Carrington* (1826-1875)	Astronomy	1860	Brewery Proprietor	Cambridge	5'9"	22 1/8"	Rad	O	X	19, 61	X	X				41	X			
Henry J. Carter* (1813-1895)	Biology (Geology)	1859		London École de Med.	5'8 1/2"	22 1/2"	L	ChE		24		X		67			X			X
Arthur Cayley* (1821-1895)	Mathematics	1852	Russia Merchant	Cambridge	5'7 1/2"	23"			X	8			X	19					X	
James Challis (1803-1882)	Astronomy	1848																		
Jacob A.L. Clarke (1817-1880)	Medicine	1854																		
Robert B. Clifton (1836-1921)	Physics	1868																		
Thomas Cobbold* (1828-1886)	Helminthology	1864	Clergy	Edinburgh	5'7"	21"			52	19	X	X		44		23	X	23, 34		X
Frederick Currey* (1819-1881)	Mycology	1858	Parliamentary Clerk	Cambridge	5'10 1/2"	21 3/4"	O(C)	ChE	X		44	X		72			X			
Charles Darwin* (1809-1882)	Biology	1834	Physician	Cambridge	6'0"	22 1/4"	L	O.ChE	1	40	26		X	41		37	X	46, 6	X	9
Thomas Davidson (1817-1885)	Biology	1857																		
Henry Debus* (1824-1916)	Chemistry	1861		Marburg	5'8 1/2"	23"							X				X			
Warren De La Rue* (1815-1889)	Astronomy	1850	Paper Manufacturer	École Saintes Barbes, Paris	5'4"	22 1/3"	L	ChE	X		X	X	X	5?		38	X	3	X	3
Duke of Devonshire (1808-1891)	Biology	1829																		
Peter M. Duncan (1821-1891)	Biology	1868				22 1/4"														
Philip Grey-Egerton* (1806-1881)	Biology	1831	Clergy	Oxford	5'9 1/2"	22"	C	ChE	46	6			X	36			25			
Earl of Enniskillen (1807-1886)	Icthyology	1829																		

	Scientific Field	Date F.R.S.	Father's Occupation	Education	Stature	Head Circumference	Political Affiliation	Religious Affiliation	Business Habits (pp. 104-107)	Education (pp. 234-260)	Energy (pp. 78-98)	Health (pp. 90-102)	Independence (pp. 121-124)	Orig. Taste Science (pp. 109-186)	Mechanical Aptitude (pp. 124-126)	Memory (pp. 107-121)	Mental Peculiarities	Religious Bias (pp. 126-191)	Scholarliness	Steadiness (pp. 103-104)
Frederick J. Evans* (1815-1885)	Physical Geog	1862	Master, RN	None after age 13	5'7 1/2"	22 7/16"			X	9, 22	X		X		3		10	X	25	X	X
John Evans* (1823-1908)	(Archaeology) Geology	1864	Teacher	Market Bosworth G. Sch.	5'10"	22 1/2"	LC			6	X		X	X	[32]	3	3	X			
William Fairbairn* (1789-1874)	Engineering	1850	Farm Servant	Millwright Apprentice	6'0"		LC	Ch. Scot.		X	23		X	X		28?	X				11
William Farr (1807-1883)	Statistics	1855																		
Arthur Farre* (1811-1887)	Medicine	1839	Physician	Cambridge	5'7"	22 1/2"			4				X			5		X			
James Fergusson (1832-1907)	India Governm Statistics	1863				22 3/4"														
William Fergusson* (1808-1877)	Medicine	1848		Edinburgh	6'0"	22"	C	ChE		42	27		X		77	6	21	X		X	X
William H. Flower* (1831-1899)	Biology (Anthropology)	1864	Author Brewer	London	6'0"	22 1/2"				·			X								
David Forbes* (1828-1876)	Geology	1858		Edinburgh	5'8 1/2"	20 1/2"	O	ChE		20, 41	1		X	X	38		15?	X	26	X	X
George C. Foster* (1835-1919)	Physical Chemistry	1869		London, Ghent, Paris, Heidelb.	5'11 1/2"	22 3/4"							X		[4]			X			
Michael Foster (1836-1907)	Physiology	1872	Physician	London																
Wilson Fox* (1831-1887)	Medicine	1870	Manufacturer	Edin., Paris, Vienna, Berlin	5'11"					21, 80	X		X				32	X			X
Edward Frankland (1825-1899)	Chemistry	1853																		
Douglas Galton* (1822-1899)	Sanitar. Engin.	1859	Independent	Roy. Military Academy	5'10 3/4"								X					X			
Francis Galton (1822-1911)	Anthropology	1860			5'9"	21 1/8"														
John P. Gassiot (1797-1877)	Astronomy	1840				24 7/8"														
Archibald Geike (1835-1925)	Geology	1865																		
Joseph H. Gilbert (1817-1901)	Chemistry	1860																		
John Hall Gladstone* (1827-1902)	Chemistry	1853	Draper	London	5'6 1/2"	22 1/2"	L	Evang ChE		85			X	X	24			X	21	
James Glaisher (1848-1928)	Mathematics	1849																		
Robert Godwin-Austen (1808-1884)	Geology (Geography)	1849																		
Robert Grant (1814-1892)	Astronomy	1865																		
John Edward Gray* (1800-1875)	Zoology	1832	Pharmacist	Middlesex & St. Barth. Hospital	5'10"	23"	L	ChE	X	X	X		X	X	40	7	X	X		X	7
Robert H. Greg* (1795-1875)	Statistics	---	Millowner	Edinburgh	5'5 1/2"	22 1/2"	LC	Unitarian					X			13		X			
William Grove* (1811-1896)	Physics	1840	Magistrate	Oxford	6'0"	24"	O(L-C)		6				X		12			X			

	Scientific Field	Date F.R.S.	Father's Occupation	Education	Stature	Head Circumference	Political Affiliation	Religious Affiliation	Business Habits (pp. 104-107)	Education (pp. 234-260)	Energy (pp. 78-98)	Health (pp. 99-102)	Independence (pp. 121-124)	Orig. Taste Science (pp. 149-186)	Mechanical Aptitude (pp. 124-126)	Memory (pp. 107-121)	Mental Peculiarities	Religious Bias (pp. 126-141)	Scholarliness	Steadiness (pp. 103-104)
Albert Günther* (1830-1914)	Zoology	1867	Bursar of Estates	Tubingen	5'8 1/2"	23 1/3"			X		3	X	X	58		17	X	52	X	X
William Guy* (1810-1885)	Statistics	1866	Physician	Cambridge	5'10 1/2"	22 1/4"	O	ChE		43	X	X	X	83		42	X		X	
Augustus Harcourt* (1834-1919)	Chemistry	1868	Admiral	Oxford	5'6 1/2"	22 1/4"	L	ChE		51	11			21		35	X		X	
Robert Harkness (1816-1878)	Geology	1856																		
William P. Hatherly* (1801-1881)	Statistics	1836	Political Reformer	Geneva	5'9"	24"	L	ChE		X		X		[82?]				X		
Samuel Haughton* (1821-1897)	(Mathematics) Geology	1858	Flour Merch. Reformer	Dublin	5'8"	21 3/4"				2	31	X	X	2			X	X	22?	X
J. Hawkshaw (1811-1891)	Engineering	1855																		
John S. Henslow* (1796-1861)	Botany (Geology)	---	Solicitor	Cambridge	5'6 3/4"							X		[70]	8		X			
James Heywood* (1810-1897)	Statistics	1839		Cambridge	5'11"	23"				47			X	[86]			X			
Rowland Hill* (1795-1879)	Statistics	1857	Teacher	Hilltop School Birmingham	5'7"	23 7/16"			X	7,31	14	X	6	85	14	8	X		X	
John R. Hind (1803-1875)	Astronomy	1851 1863																		
Thomas Kirst* (1830-1892)	Mathematics	1861	Woolstapler	Marburg Göttingen	6'2 1/2"	24"	O	O	X	X	X	X	X	17			X	X	X	2
Henry Holland* (1788-1873)	Medicine	1815	Physician	Edinburgh																
Joseph D. Hooker* (1817-1911)	Botany	1847	Botanist	Glasgow	5'10 1/2"	21 1/2"				16,55	2	X	X	64		18	X		X	X
Richard M.M. Houghton (1809-1885)	Statistics	1868																		
William Huggins (1824-1910)	Astronomy	1865				22"														
George Humphry* (1820-1896)	Medicine	1859	Lawyer	St. Bartholomew's Hosp.	5'11"	22"				3,30		X		76						
Thomas H. Huxley* (1825-1895)	Zoology	1851	Teacher	London	5'10"	23"			3			X		52	9			X	29	
Henry James (1803-1877)	Geology	1848				22 3/8"														
William Jardine (1800-1874)	Biology	1860																		
John G. Jeffreys* (1809-1885)	Zoology	1840	Solicitor	Swansea Gram. School	5'8 1/2"	23"	C	ChE	X	54		X		[51?]			X		X	X
Fleeming Jenkin* (1833-1885)	Engineering	1865	Captain RN	Edin. Academy, Paris, Genoa	5'7 3/4"	21 3/4"	L	ChE		13		X		88			X			
William Jenner (1815-1898)	Medicine	1864																		
William Stanley Jevons (1835-1882)	Statistics	1872	Iron Merchant	London	5'5 1/2"	23"	L	Unitarian		4,14 48	43	X		84	15		X			
T. Rupert Jones* (1819-1911)	Geology	1872	Silk Merchant	Medical Apprentice	5'3"	23"	L	ChE		58		X		33						

	Scientific Field	Date F.R.S.	Father's Occupation	Education	Stature	Head Circumference	Political Affiliation	Religious Affiliation	Business Habits (pp. 106-107)	Education (pp. 234-260)	Energy (pp. 78-98)	Health (pp. 99-102)	Independence (pp. 121-124)	Orig. Taste Science (pp. 149-186)	Mechanical Aptitude (pp. 124-126)	Memory (pp. 107-121)	Mental Peculiarities	Religious Bias (pp. 126-141)	Scholarliness	Steadiness (pp. 103-104)
James P. Joule (1818-1889)	Physics	1850																		
Robert J. Kane (1807-1890)	Chemistry	1849																		
Francis Kiernan (1800-1874)	Physiology	1834																		
William Lassell* (1799-1880)	Astronomy	1849	Merchant Apprentice		5'9 1/2"	22 1/2"	L	Congr	X	X	X	X	X	1						X
John B. Lawes* (1814-1900)	(Agriculture) Chemistry	1854	Landowner	Oxford	5'8"		Rad	ChE	X	X	X	X				2	X	X	X	5
Robert Lee (1793-1877)	(Physiology) Medicine	1830																		
John G. Lefevre (1797-1879)	Medicine	1820						25"												
Humphrey Lloyd (1800-1881)	Astronomy	1836																		
Joseph N. Lockyer (1836-1920)	Astronomy	1869				22 1/8"														
John Lubbock* (1834-1913)	Zoology (Anthropology)	1858	Banker	Eton	5'10"	21"	L	ChE	X	45	X	X		42				X		
Charles Lyell (1797-1875)	Geology	1826																		
Robert Main* (1800-1878)	Astronomy	1860		Cambridge	5'6"	23"				1	X	X	X	(15?)				X	11	1
John Marshall* (1818-1891)	Medicine	1857	Solicitor	London	5'10"	23"	C	ChE or Scot		X								X	27	
N. Story-Maskelyne* (1823-1911)	Mineralogy	1870	A.K. Story, FRS	Oxford	5'7 3/4"	23 1/16"				X				(39?)				X		
James C. Maxwell* (1831-1879)	Physics	1861	Landowner	Cambridge	5'8"	23 1/2"				5,18,25		5	X	(16)				X	10	4
Charles Merrifield* (1827-1884)	(Mathematics) Naval Engin.	1863		No College Education	5'4 1/4"	23"	L	ChE	X	35	24	X	9?	91?				X	X	X
John Miers* (1789-1879)	Botany	1843	Jeweler Mineralogist	No College Education						X				(68)				X		
William H. Miller* (1801-1880)	(Mineralogy) Chemistry	1838		Cambridge	5'5"	22 1/4"	C	ChE		72										
George Mivart* (1827-1900)	Anatomy	1869	Hotel Owner	London St. Mary's	6'0"			RC	X				X	45				X		
William Newmarch* (1820-1882)	Statistics	1861		No formal Education	5'11"								X					X		
Alfred Newton* (1829-1907)	Ornithology	1870	Statesman	Cambridge	6'0"	22 1/4"				11,78	32	3		48						
William Odling (1829-1921)	Chemistry	1859																		
Sherard Osborn* (1822-1875)	Geography	1870	Army	Navy	5'8 1/2"	24"	L	ChE	X	77	10	X	X	X		X	X	X	X	X
Richard Owen* (1804-1892)	Comp. Anat.	1834	West India Merchant	Edinburgh	5'11"	23" (FG)				27,32 50	15		X	43		34?		X		
James Paget* (1814-1899)	Medicine	1851	Brewer Shipowner	St. Bartholomew's Hosp.	5'10 1/4"				X	X	X	X		74				X	X	X

	Scientific Field	Date F.R.S.	Father's Occupation	Education	Stature	Head Circumference	Political Affiliation	Religious Affiliation	Business Habits (pp. 104-107)	Education (pp. 234-260)	Energy (pp. 78-98)	Health (pp. 99-102)	Independence (pp. 121-124)	Orig. Taste Science (pp. 149-186)	Mechanical Aptitude (pp. 124-126)	Memory (pp. 107-121)	Mental Peculiarities	Religious Bias (pp. 126-141)	Scholarliness	Steadiness (pp. 103-104)
William K. Parker* (1823-1876)	Compar. Anat.	1865	Farmer	Charing Cross Hospital	5'9 1/2"	23"	O.C	Wesley ChE	X	33	X			47			X	18	X	8
Edmund A. Parkes* (1819-1876)	Medicine	1861	Landowner	London	5'8"	21 1/2"	O	OChE	81	17	X	X		60		X	X			
Frederick W. Pavy (1829-1911)	Medicine	1863																		
John Percy (1817-1889)	Chemistry	1847																		
James B. Pettigrew (1834-1908)	Physiology	1868																		
John Phillips* (1800-1874)	Geology	1834	Excise Off.	Wiltshire Gram. School	5'7 1/2"	22 5/8"							[2]				X			
Lyon Playfair* (1818-1898)	Chemistry	1848	Bengal Hosp. Inspector	Glasgow, Edin. Giessen	5'8 1/2"	22 1/2"	L	Presb. ChE	X	15,84	X	X	X	27	2		X	X	X	
William Pole* (1814-1900)	Engineering (Music)	1861		Engineer's Apprentice	5'10"	22 1/2"	O	Cong. ChE	7	X	41	X	X	90			X		X	X
Joseph Prestwich* (1812-1896)	Geology	1853	Wine Merchant	London	5'10 1/2"	22 3/4"	L	ChE	60,73 12	37	X	8	37				X	5	X	X
Bartholomew Price (1818-1898)	Mathematics	1852																		
Charles Pritchard (1808-1893)	Astronomy	1840																		
Andrew C. Ramsay* (1814-1891)	Geology	1849	Manufacturing Chemist	No College Education	5'8"	22 1/2"														
Henry C. Rawlinson (1810-1895)	Archaeology	1850																		
John Rennie (1794-1874)	Engineering	1823																		
George H. Richards (1820-1896)	Naval Engin.	1866				21 1/4"														
Benjamin W. Richardson (1828-1896)	Medicine	1867																		
Thomas Robinson (1792-1882)	Astronomy (Math. Phys.)	1856																		
George Rolleston (1829-1881)	Medicine	1862																		
Henry E. Roscoe* (1833-1915)	Chemistry	1863	Judge	London Heidelberg	5'11"	23 1/2"	L	Unitarian	H,10 28	X	X			23	.					
Laurence P. Rosse* (1840-1908)	Astronomy	1867	Astronomer Landowner	Dublin	5'7 1/2"	21"								X		[11?]		X		
John S. Russell (1808-1882)	Civil Engineering	1849																		
Edward Sabine* (1788-1883)	Astronomy	1818		Royal Military Academy		21 3/4"				·				X		[14?]				
Robert Salisbury* (1830-1903)	Physics	1869	Marquis of Salisbury	Oxford	6'2"	23"	C	ChE			44, 82	X	X							
George Salmon (1819-1904)	Mathematics	1863																		
J. Burdon-Sanderson* (1828-1905)	Physiology	1867		Edinburgh Paris	6'2"	23"							X			10	X			

	Scientific Field	Date F.R.S.	Father's Occupation	Education	Stature	Head Circumference	Political Affiliation	Religious Affiliation	Business Habits (pp. 104-107)	Education (pp. 234-260)	Energy (pp. 78-98)	Health (pp. 99-102)	Independence (pp. 121-124)	Orig. Taste Science (pp. 109-186)	Mechanical Aptitude (pp. 124-126)	Memory (pp. 107-121)	Mental Peculiarities	Religious Bias (pp. 126-141)	Scholarliness	Steadiness (pp. 103-106)
Philip L. Sclater* (1829-1913)	Ornithology	1861	Landowner	Oxford	5'11"	23"	L	ChE	2	X	25	X		59			X			
Robert H. Scott* (1833-1916)	(Mathematics) Meteorology	1870		Dublin	5'11"	24"	C	ChE	X	83	30	X	X	8		30	X		X	X
William Sharpey (1802-1880)	Physiology	1839				24 3/8"			.											
Francis Sibson (1814-1876)	Medicine	1849				23"														
William Siemens* (1823-1883)	Engineering	1862	Farmer	Magdeburg Göttingen	5'10 1/2"	23 3/8"	L	Prot.	.	X	X	X	X			14	X		X	X
John Simon (1816-1904)	Medicine	1845																		
Archibald Smith* (1813-1872)	Mathematics	1856	Author Geologist	Glasgow Cambridge		23 1/2"						X		19			X			
Edward Smith (1818-1874)	Medicine	1860																		
Henry J. Smith (1826-1883)	Mathematics	1861				22 1/8"														
Piazzi Smyth* (1819-1900)	Astronomy	1857	Admiral	No College Education	5'9"	23 1/2"		Bible Xtn				X	4	[9]			X			
Warington Smyth* (1817-1890)	Mineralogy	1858	Admiral	Cambridge	5'7 1/2"	21 5/8"				79, 57	0	X	X	35		X	X			
William Smythe (1816-1887)	Military Engineering	1864																		
Herbert Spencer* (1820-1903)	Biology (Philosophy)	---	Schoolmaster	No College Education	5'10"							X					X			
William Spottiswoode* (1825-1883)	Mathematics	1853	Printer	Oxford	5'11 1/2"	22 3/4"			X	X	X	X		[20?]			X			
Earl Stanhope (1805-1875)	History	1827																		
John Stenhouse* (1809-1880)	Chemistry	1848	Calico Printer	Glasgow Giessen	5'8 1/2"	23"	L	U. Pres.		23, 53		X								
Balfour Stewart* (1808-1887)	Physics	1862	Tea Merchant	Edinburgh	5'10 1/4"	22"	LC	ChE				X	X	1	7			X	8	
George G. Stokes (1819-1903)	Physics	1851																		
William Stokes* (1804-1878)	Medicine	1861	Prof. of Medicine	Edinburgh	5'10 1/2"	24"			X	37, 74	X	4	X	56		X	X	14	X	X
Richard Strachey* (1817-1908)	Geography	1854	India Civil Service	Addiscombe Milit. Acad.	5'8 1/2"	22 3/8"						X		[13]			X	2		
Alexander Strange (1818-1876)	Geography	1864																		
Paul E. de Strzelecki (1796-1873)	Geography	1853																		
James J. Sylvester (1814-1897)	Mathematics	1839				23 1/8" (FG)														
Christopher R. Talbot (1803-1890)		1831	Lord Lt. of Glamorgan, M.P.																	
William Fox Talbot (1800-1877)	Mathematics (Photography)	1831																		

Name	Scientific Field	Date F.R.S.	Father's Occupation	Education	Stature	Head Circumference	Political Affiliation	Religious Affiliation	Business Habits (pp. 104-107)	Education (pp. 230-260)	Energy (pp. 78-98)	Health (pp. 99-102)	Independence (pp. 121-124)	Orig., Taste Science (pp. 149-186)	Mechanical Aptitude (pp. 124-126)	Memory (pp. 107-121)	Mental Peculiarities	Religious Bias (pp. 126-141)	Scholarliness	Steadiness (pp. 103-104)
Allen Thomson* (1809-1894)	Biology	1848	Physician	Edinburgh	5'7"	22 1/8"			X	40, 33	X	X	X	57	12	X	X		X	X
Thomas Thomson (1817-1878)	Botany	1855																		
William Thomson (1824-1907)	Physics	1851																		
Isaac Todhunter (1820-1884)	Mathematics	1862																		
Edward B. Tylor (1832-1917)	Anthropology	1871					22"														
John Tyndall (1820-1893)	Physics	1852																		
Charles B. Vignoles (1793-1875)	Railway Engineering	1855																		
John Voelcker* (1822-1884)	Agricultural Chemistry	1870	German Merchant	Gottingen Giessen	5'9"	23"	L	Bap-tist		X		X	X	25						
Arthur Walden (1824-1878)	Zoology	1871																		
Thomas Watson* (1792-1882)	Medicine	1859		Cambridge	5'10 1/2"	22 1/2"	C	ChE		X	21	X		79						X
Andrew S. Waugh (1810-1878)	Military Engineering	1858																		
Thomas Webster (1810-1875)	Physics	1847																		
Charles Wheatstone (1802-1875)	Physics	1836																		
Alexander Williamson* (1834-1904)	Chemistry	1855	India House Clerk	Heidelberg Geissen	6'0"	23"							X				X			
William Williamson* (1824-1904)	Biology	1854	Gardner	Little formal Education	5'7 1/2"	22 7/8"	L	Wes-ley	X	56	18	X	X	55		X	X	10	X	X
Robert Willis (1800-1875)	Mathematics	1830																		
Philip J. Yorke* (1799-1874)	Chemistry	1849	Prebendary of Ely	Harrow	5'10"	22"				X		X								

INDEX TABLE ERRATA

"David G. Ansted" should be "David *T*. Ansted"

"Duke of Argyll (1826–1900)" should be "Duke of Argyll (*1823*–1900)"

"x" in column under "steadiness" for Henry J. Carter should be "*7*"

"Farm Servant" under "father's occupation" for William Fairbairn should be "*Farm-baliff.*"

"James Fergusson (1832–1900)/India Governm. Stat." should be "James Fergusson (*1808–1886*)/*Archaeology (Architecture)*"

"John P. Gassiot/Astronomy" should be "John P. Gassiot/*Physics*"

"James P. Glaisher (1848–1893)/Mathematics" should be "James P. Glaisher (*1809–1903*)/*Astronomy*"

"7" in column under "steadiness" for John E. Gray should be "*x*"

"Robert Main (1800–1878)" should be "Robert Main (*1808*–1878)"

"William K. Parker (1823–1876)" should be "William K. Parker (1823–*1890*)"

"Balfour Stewart (1808–1887)" should be "Balfour Stewart (*1828*–1887)"

"Allen Thomson (1809–1894)" should be "Allen Thomson (1809–*1884*)

"Alexander Williamson (1834–1904)" should be "Alexander Williamson (*1824*–1904)"

"William Williamson (1824–1904)" should be "William Williamson (*1816–1895*)"